Chief Officer

Principles and Practice

SECOND EDITION

David J. Purchase, EFO
Fire Chief (Retired)
Norton Shores Fire Department
Norton Shores, Michigan

World Headquarters
Jones & Bartlett Learning
5 Wall Street
Burlington, MA 01803
978-443-5000
info@jblearning.com
www.jblearning.com

International Association of Fire Chiefs
4025 Fair Ridge Drive
Fairfax, VA 22033
www.IAFC.org

National Fire Protection Association
1 Batterymarch Park
Quincy, MA 02169-7471
www.NFPA.org

Jones & Bartlett Learning books and products are available through most bookstores and online booksellers. To contact Jones & Bartlett Learning directly, call 800-832-0034, fax 978-443-8000, or visit our website, www.jblearning.com.

Substantial discounts on bulk quantities of Jones & Bartlett Learning publications are available to corporations, professional associations, and other qualified organizations. For details and specific discount information, contact the special sales department at Jones & Bartlett Learning via the above contact information or send an email to specialsales@jblearning.com.

10423-3

Production Credits

Chief Executive Officer: Ty Field
Chief Product Officer: Eduardo Moura
VP, Executive Publisher: Kimberly Brophy
VP, Sales, Public Safety Group: Matthew Maniscalco
Director of Sales, Public Safety Group: Patricia Einstein
Executive Editor: William Larkin
Senior Acquisitions Editor: Janet Maker
Associate Managing Editor: Amanda Brandt
Associate Director of Production: Jenny L. Corriveau
Production Editor: Lori Mortimer
Senior Marketing Manager: Brian Rooney
Production Services Manager: Colleen Lamy
Senior Production Specialist: Carolyn Downer
VP, Manufacturing and Inventory Control: Therese Connell
Composition: diacriTech
Cover Design: Kristin E. Parker
Associate Director of Rights & Media: Joanna Lundeen
Rights & Media Specialist: Robert Boder
Media Development Editor: Shannon Sheehan
Cover Image: © Cindy Jurkas, Realistic Photography. Used with permission.
Printing and Binding: LSC Communications
Cover Printing: LSC Communications

Library of Congress Cataloging-in-Publication Data
Purchase, David (Firefighter), author.
Chief officer : principles and practice / David Purchase.—Second edition.
pages cm
"International Association of Fire Chiefs, National Fire Protection Agency."
Includes index.
ISBN 978-1-284-03842-2
1. Fire departments—Administration. 2. Fire chiefs. I. Title.
TH9158.P87 2017
363.37068'4–dc23
2015034854

6048

Printed in the United States of America
21 20 19 10 9 8 7 6 5

Brief Contents

Contents

Instructor Resources

Instructor Resources

Instructor's ToolKit CD

Preparing for class is easy with these resources. To meet your course delivery needs, the resources in the Instructor's ToolKit (ITK) are available for Fire Officer III, Fire Officer IV, or Fire Officer III and IV (combined). The ITK includes:

- **Presentations in PowerPoint format** provide you with a powerful way to make presentations that are educational and engaging to your students. These slides can be modified and edited to meet your needs.
- **Lesson Plans** provide you with complete, ready-to-use lesson plans that include all of the topics covered in the text. Offered as Word documents, the lesson plans can be modified and customized to fit your course.
- **Test Bank** contains multiple-choice questions and allows you to create tailor-made classroom tests and quizzes quickly and easily by selecting, editing, organizing, and printing a test along with an answer key that includes page references to the text.
- **Image and Table Banks** provide you with a selection of the most important images and tables found in the textbook. You can use them to incorporate more images into the PowerPoint presentations, make handouts, or enlarge a specific image for further discussion.
- **Chief Officer in Action Answers** identify the correct answers to the end-of-chapter multiple-choice questions. These answers can facilitate class discussion or activity correction.

Acknowledgments

Author

David J. Purchase, EFO

Chief Purchase served 38 years in the fire service, retiring in 2013 after 14 years as fire chief for the city of Norton Shores, where he resides with his wife, Kari. He has been a certified fire instructor for 29 years and is currently an Instructor II, having taught over 400 fire officer programs around the state of Michigan since 1996. An EFO graduate of the National Fire Academy, Chief Purchase has received various recognitions throughout his career, including the Michigan Jaycees, State Firefighter of the Year, Fire Service Instructor of the Year presented by the Michigan Fire Service Instructors Association and the Michigan Association Fire Chiefs, and Fire Chief of the Year. He served for 10 years as a governor appointee on the Michigan Firefighters Training Council. He is a past president of both the Western Michigan Association of Fire Chiefs and the Michigan Association of Fire Chiefs and past vice president of the Michigan Fire Service Instructors Association. Chief Purchase has presented programs at fire service conferences around the country, including VCOS, FDIC, FDIC East, Ohio State Fire School, Michigan and Wisconsin State Fire Chiefs, and Fire House Expo. He has been a contributing author and/or editorial board member for Jones & Bartlett Learning publications, including *Chief Fire Officer's Desk Reference* (Training chapter), *Fire Service Instructor* and *Hazardous Materials* curriculum development, and *Chief Officer*.

First Edition Authors

Jackson Baynard, MHRM
Lieutenant
Henrico County Division of Fire
Henrico, Virginia

Harry R. Carter, MIFireE, CFO, PhD
Chairman, Board of Fire Directors, Howell Township Fire District # 2
Battalion Chief (Retired), Newark Fire Department
Newark, New Jersey

Riley Caton, BS, EFO
Fire Chief (Retired)
Gresham Fire and Emergency Services
Gresham, Oregon

John Fennell
General Counsel (Retired), Office of the State Fire Marshal
Chicago, Illinois
Fire Chief (Retired), Elmhurst Fire Department
Elmhurst, Illinois

Ken Folisi, MS
Assistant Professor, Fire Science Management, Lewis University
Romeoville, Illinois
Battalion Chief (Retired), Lisle Woodridge Fire District
Lisle, Illinois

Richard B. Gasaway, PhD, EFO, CFO, MICP
Executive Director, Center for the Advancement of Situational Awareness and Decision Making
Chief Scientist, Public Safety Laboratory
President, Gasaway Consulting Group
Minneapolis, Minnesota

James M. Grady III, BA, CFO
Chief
Frankfort Fire Protection District
Frankfort, Illinois

Shawn Kelley
Director of Strategic Services/GPSS
International Association of Fire Chiefs
Fairfax, Virginia

Barry McLamb, BS, EFO
Battalion Chief
Chapel Hill Fire Department
Chapel Hill, North Carolina

David R. Peterson, EFO
Fire Chief
Plainfield Fire Department
Plainfield, Michigan

David J. Purchase, EFO
Fire Chief (Retired)
Norton Shores Fire Department
Norton Shores, Michigan

Paul Ricci, EFO, CFO
Fire Chief
Sandusky Fire Department
Sandusky, Ohio

John Rukavina, JD, CFO, FIFireE
Director
Public Fire Safety Services
Asheville, North Carolina

Adam K. Thiel, MA, CEM, CFO, MIFireE
Fire Chief
City of Alexandria Fire Department
Alexandria, Virginia

Mitchell Waite, PhD, CFO
Fire Chief, Wisconsin Rapids Fire Department
Lieutenant Colonel, United States Army Reserve
Wisconsin Rapids, Wisconsin

Michael J. Ward, MGA, MIFireE
Senior Consultant
Fitch and Associates
Platte City, Missouri

4. Objectives: These are the tasks that must be accomplished if an organization is to meet its goals and fulfill its mission. The terms *goal* and *objective* are often used interchangeably, and tasks and subtasks should align with the goal or objective. Once the objectives are created, it is important to assign a committee or individual to be accountable for completing these tasks. This allows the leader to track progress by discussing the assigned task with the individual or group. A timeline is also important so employees know how long they have to complete tasks. If the time to complete the project is insufficient, the effective leader will need to provide more resources to complete the task on time or reevaluate the task and adjust the timeline.

For a fire department to be successful, its leaders must understand their role in the creation and fulfillment of these elements. Leading a fire/rescue/emergency medical service (EMS) department usually takes place under conditions that are less than optimal, and the individuals being led are formed into groups. Fire departments are most successful when their members operate as a team in a coordinated effort (Paulsgrove 2003). These groups of individuals function best when led by an effective leader.

The development of effective leaders, who are able to command respect and loyalty as well as function effectively in crisis situations, is a necessary precursor to the delivery of safe and efficient firefighting operations. Teams need leaders who can motivate those who labor under them as well as coordinate and interface well with those to whom they report. Establishing the right attitude creates an atmosphere that enables the best response from team members. Effective leaders also understand that hiring the right people (who possess the correct skill sets for a career in firefighting) and promoting the organization's best and brightest are important variables in an organization's success (see the "Human Resources" chapter).

Structure

The most basic organizational principle for a fire department is the division of work among the operating units and their individual members. This division of work is based on specific functions within the department (e.g., operations, fire prevention, training) and is structured to avoid overwhelming one individual or group with excessive responsibility.

The organization of fire departments varies depending on the size and population of the communities they serve. A small department usually has a simple organizational structure that allows frequent personal contact among members, which facilitates coordination of activities. In larger departments, such contact is not feasible, and an efficient structure that provides effective communications becomes increasingly important. The important concepts of unity of command and span of control can assist leaders in facilitating communication and understanding. Sample organizational structures for small and large departments can be found in FIGURE 1-2.

Culture

Organizational culture is the beliefs, values, and norms shared by the members of an organization. These beliefs, values, and norms generally go undocumented—learning them is part of the rite of passage of members new to the organization. In fact, some believe it is taboo to openly discuss some cultural elements. Most importantly, organizational culture exists parallel to—never as part of—the official organizational structure.

Common organizational culture elements include the following:

- *Shared meanings*: Common interpretation of events (e.g., every fire fighter who also handles EMS responses knows what *frequent flyer* means)
- *Perceptions*: Agreed-on views of how the work environment and the world work
- *Behavioral codes*: Culturally acceptable behavior (e.g., if an official work shift begins at 8:00, a member who arrives 5 minutes early may be shunned if the culture dictates arrival 1 hour early for a shift)
- *Prescriptions and preferences*: The fire fighter training manual, my fire department's way, and my captain's way
- *Basic values*: Accepted mentality, importance of safety concerns (e.g., aggressive firefighting, nonchalance in the face of danger, wearing seat belts)
- *Myths, legends, heroes, and heroines*: Stories of the big one and who played what role, the long-departed fire fighter who had more fire ground credibility and leadership skill than any officer, etc.
- *Emblems*: T-shirts, tattoos, company logos
- *Rituals*: "Wet-downs" on completion of basic training, etc.

New members must adapt to the culture of their organization and find their role within it. Beyond the long-standing rituals and slang terminology, however, an organizational culture can emerge from new priorities and perspectives from the fire service industry.

For example, the fire service aims to make significant progress in reducing the number of fire fighter fatalities and injuries. Any progress will take cultural change driven by effective leadership. Creating a safety culture within a department is everyone's responsibility, not just the chief officer's. There is a huge difference between talking about safety practices and actually creating, implementing, and institutionalizing a safety culture. True leaders understand that this is something they owe to their fire fighters and their families. The very necessity of a safe operational scene must have its roots in the establishment of a safe work environment in the station and on the drill ground. Leaders who allow their fire fighters to skirt basic safety practices in these areas cannot expect those same fire fighters to suddenly become safe employees on the incident scene. Safe habits learned are safe habits practiced.

Life Safety Initiatives

1. Define and advocate the need for a cultural change within the fire service relating to safety; incorporating leadership, management, supervision, accountability, and personal responsibility.
4. All fire fighters must be empowered to stop unsafe practices.

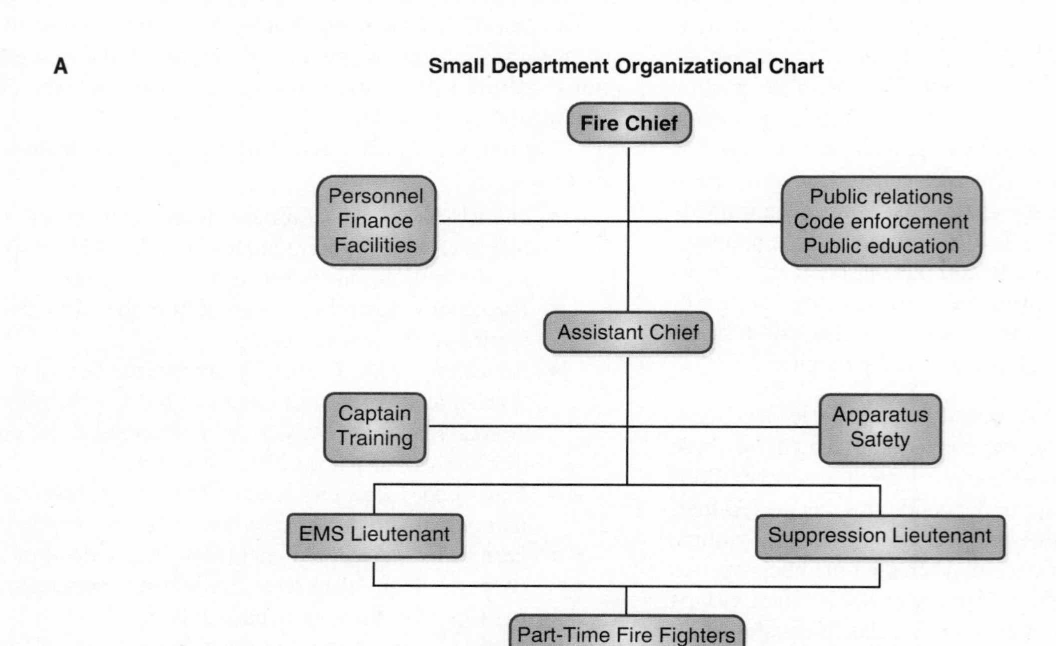

FIGURE 1-2 **A.** Sample fire department structure: small department.

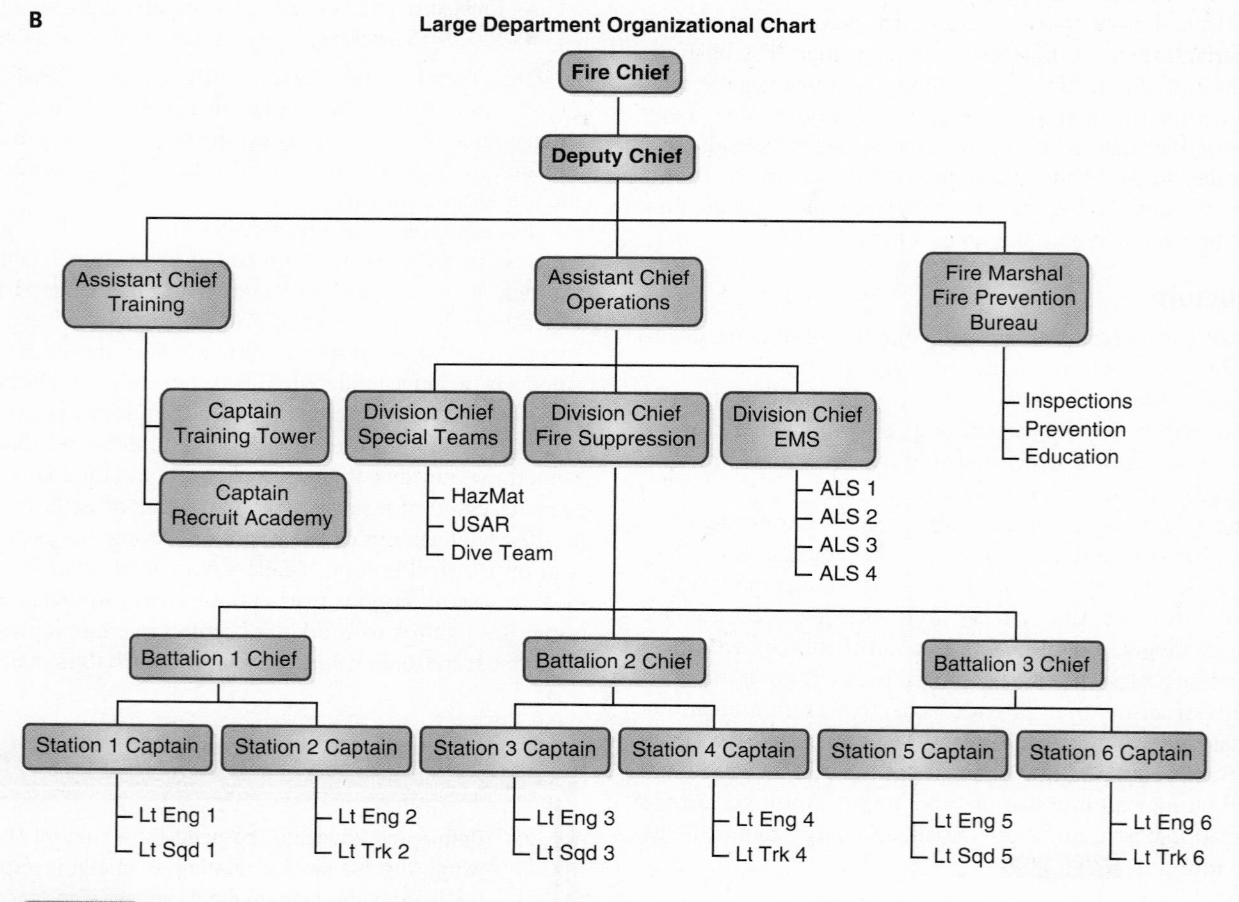

FIGURE 1-2 **B.** Sample fire department structure: large department.

Trends

The search for more effective and efficient organizations continues and has followed several trends, including regionalization and flexibility.

Regionalization

The data revolution and internationalization of manufacturing technologies have helped globalize enterprises, resulting in substantial changes in organizations and increasingly complex organizational cultures.

The fire service equivalent of globalization is regionalization. This ranges from the sharing of specific services or facilities to consolidation of fire departments across regions and has been driven largely by interests in more efficient, less costly operations. The impact of regionalization (or consolidation) on an organization lies more in the collision of organizational cultures than on the functions of the members. An excellent example is the consolidation of fire and EMS organizations in large cities in which the fire and EMS organizational cultures had previously been independent.

Flat Organizations

The search for flatter organizations—organizations with fewer layers of supervision and management and more front-line workers—continues, particularly in an environment in which a local, regional, or national economy's ability to pay is outstripped by the cost of the organizational product or service. In labor-intensive organizations like fire departments, significant reductions in costs can often be achieved by reorganizing supervisor and manager positions out of the bureaucracy or by reducing the overall size of the bureaucracy at all levels.

A variation of the flatter organization is the slimmer organization—an organization that sheds costs by reducing or eliminating certain outputs or services. In fire service organizations, this is often an examination of response-reduction options (i.e., reducing responses to automatic alarms or minor medical emergencies) or reduction or elimination of some organization functions (like public fire safety education) in the hope that such reductions can result in a substantial reduction in costs.

Flexibility

A bureaucratic hierarchy is viewed by many as a rigid, inflexible system that cannot effectively respond to changes in demand for service, technology, or the environment. In the interests of effectiveness and efficiency, flexibility has become a high priority. One reflection of this interest is the adoption of processes and procedures that enable an organization to respond gracefully to many varied situations. The assumption is that a combination of fewer rules and procedures with more training and individual autonomy will enable an organization to shift gears with greater ease to meet new demands and challenges.

An example of an introduction of flexibility to fire service organizations is the development and adoption of the NIMS incident command systems (ICS). A major operating principle of the ICS is that the ICS framework can be applied to any emergency situation; that is, with appropriate training and equipment, fire fighters should be as capable of responding successfully to a terrorism emergency as they are to a structure fire. A second principle is that an ICS should support an integrated emergency response in which several organizations—fire, law enforcement, and public health—can depend on and work closely with each other, yielding a successful outcome.

The fire service is still struggling with the relationship between the security of standard operating procedures (SOPs) and flexibility. The greatest concern was to remove the contradiction between training fire fighters to act swiftly, surely, and on their own in the face of danger, while at the same time training them to take orders unhesitatingly when working under command.

Hundreds of books have been written on the nature of organizations, and hundreds more on effective management and leadership of organizations. Chief fire officers must, at a minimum, take organizational structures and culture into account in setting and achieving their goals; at best, they should take advantage of organizational structure and culture as resources for organizational growth.

Effective Leadership in the Fire Service

Every successful organization has a common denominator: an effective form of leadership. Just being a good leader is a start, but goals must be set and objectives, plans, and orders created to allow for the successful operation of the agency. This requirement exists in a fire department just like it does in any other organization, making up the elements of structure, process, and behavior. These elements form the framework of the organization, within which the leader will operate.

To begin a discussion on the various leadership options, we assume that all of these models are based on the premise that some type of hierarchical leadership is needed in formal organizations. A leadership plan for the fire service can be developed by combining various aspects of the different leadership theories into an approach that best typifies leadership demands FIGURE 1-3. Beyond that, the issue of people-oriented versus task-oriented leadership is explored because it affects any model for fire service leadership.

FIGURE 1-3 Leadership in the fire service should combine aspects of different leadership theories.

Military Influence

Favreau lists a number of functions that have guided leaders during military conflicts (Favreau 1973, 28–29):

- Be technically and tactically proficient.
- Know yourself.
- Seek self-improvement.
- Know your people and look out for their welfare.
- Keep your people informed.
- Set the example.
- Train as a team.
- Seek responsibility and take responsibility for your actions.

Leadership within life-and-death situations creates similarities between war and firefighting, both of which involve:

- Situations involving life and death
- Situations in which decisions must be made quickly
- Situations in which decisions must be made without having all the facts to make the decision
- Situations in which members may be experiencing fear and stress

Within these similarities are connections that bring the military style of leadership to the fire service world. The first of these comes in the area of information available for decision-making purposes. During a firefighting operation, effective leaders must rely on those people who trust them to work diligently within the pressure-cooker environment of the fire-ground operation. The leader's ability to inspire people to labor on his or her behalf lies at the root of any success he or she will have.

Both a battlefield and a fire ground are areas of danger, excitement, and uncertainty; danger is inherent. Difficult situations can be resolved only by simple decisions and simple orders.

Fire Service Leadership Model

The preceding thoughts on leadership can be distilled into a list of attributes and skills for practicing fire service leaders and can serve chief officers well in times of uncertainty and concern TABLE 1-1.

The principles and methods of leadership cannot be taught in the classroom alone. Current leaders in any environment should aim to exemplify these attributes as they operate within their fire departments, thus influencing their followers to improve their own leadership skills.

Leadership requires integrating support and concern for people with consideration of the environment and potential tasks. The fire service leadership model would include the following attributes:

- Knowledge of emergency operations
- A firm list of tasks that must be accomplished
- A concern for people
- A thorough understanding of the mission(s) to be accomplished by operational forces
- The ability to process information quickly and accurately
- The ability to generate trust within the ranks of the people with whom the leader is working
- The ability to stimulate those people with whom the leader works to strive toward the goals of the organization and the emergency operation
- The ability to evaluate and encourage individual and team improvement
- The ability to recognize and accept change
- The ability to understand, encourage, and enforce safe behavior

In his book, *Fire Service Leadership: Theories and Practices*, Mitchell Waite highlights several examples of leadership that can lead to organizational success, using the term *leadership* as an acronym to guide chief officers in their pursuit of professional excellence. Although the concepts may seem simple enough, fire service officers have often struggled to master them.

- **L**ead from the front.
- **E**ffectively communicate.
- **A**dvise, mentor, and counsel.
- **D**ecide, act, and evaluate.
- **E**nvision the future.
- **R**emain flexible.
- **S**hare your knowledge.
- **H**onesty is always right.
- **I**nvest in your personnel.
- **P**repare to fail, if you fail to prepare.

A true leader is not afraid to get out with the troops and attempt to understand the job from their perspective. This type of leader will exemplify the adage, "Talk the talk and walk the walk." If leaders believe in what they are doing, they should have no qualms about leading the charge.

Chief officers must make decisions on a daily basis. Many of the decisions are small, some are important, and fewer are critical, but decisions have to be made nonetheless. The more difficult decisions might make some personnel unhappy, but as long as the individual does what is in the best interest of the organization, while keeping the best interest of the employees in mind, he or she can successfully lead the team.

Table 1-1 Fire Service Leadership Attributes

Integrity	Courage
Honesty	Pride
Determination	Faith
Tact	Judgment
Endurance	Initiative
Responsibility	Loyalty
Selflessness	Predictability
Self-discipline	Dependability

Delegation

For the most part, chief officers seem to have little problem with delegating tasks on an emergency scene. They are quick to recognize the importance of limiting span of control when faced with an active fire situation. They understand that they cannot fight the fire alone, nor can they directly control every supervisory task necessary on any fire ground. Why is it, then, that some of these same officers find themselves underwater

> **Chief Officer Tip**
>
> **Making Decisions in an Organizational Environment**
>
> Leaders may find themselves in either of the following types of organization: rules based or values based. In a rules-based organization, the general operational mode is that if something is not forbidden in writing, then it must be okay. Leaders must be prepared for subordinates to challenge the status quo as they attempt to shape or develop organizational policy in areas where the written standards end. In this case, leaders must learn how to guide the actions of the followers in a positive direction as they attempt to operate "outside the lines." With a values-based organization, leaders must decide which actions are acceptable based on the established values of the organization. These decisions are made whether or not a specific written rule or procedure exists. In both a rules-based organization and a values-based organization, the leader's job can be made easier if there is a department mission statement and a code of conduct/ethics established to help make decisions that may have to be judged on the basis of what is morally right.

when navigating the world of fire service administration and management?

Delegation is a necessary part of leadership. All good leaders understand that they are judged not on what they do but on how well the team performs. Delegation is an active skill that fire officers must practice. Becoming comfortable with delegating aspects of the daily routine is an important part of not only the chief officer's professional growth but of the growth of the line officers and fire fighters themselves.

In delegating tasks, chief officers must choose wisely, ensuring that the task and the individual assigned the task are a fit. It should be a participative process that allows the person being delegated the opportunity to provide his or her own input into the final outcome, but also allows the leader to monitor the process to provide any needed assistance. The level of confidence in the subordinate, on the part of the leader, to successfully complete the task will assist in determining the timing and amount of feedback requested of the subordinate on task progress. In all cases of delegation, communication is vital to a successful outcome. Subordinates must be free to accept or reject the task and must clearly understand the task parameters, limits of authority, and responsibility. The use of delegation within an organization can assist chief officers with the development of subordinates by exposing them to new processes and experiences.

Delegation is also an important component of succession planning and can help identify those candidates worthy of further mentoring, thus strengthening the organization across all levels while at the same time preparing it to move forward in the future.

Leading versus Managing

Many people, including many fire service leaders, have difficulty identifying the differences between leaders and managers. Leadership is focused on influencing people to take action to achieve a desired end state. This may be in relation to a structure fire, a military engagement, a business venture, or a sporting event. Leadership is influence, and a leader who cannot influence his or her subordinates will not be viewed as a leader. Simply put, a leader is not leading if no one is following.

Good leaders understand the science behind leadership. Great leaders understand the art of leadership and its many nuances. Great leaders can be poor managers. Conversely, good managers can be poor leaders. Managers are like staff officers in the military. They usually have no troops to influence, but rather, they manage tasks and situations. Management is more in alignment with transactional leadership, whereas leaders are more transformational in nature. However, the organization that understands and recognizes these differences can formulate outstanding teams and coalitions that will lead to great organizational success. Having an outstanding transformational leader surrounded by other transformational and transactional leaders can lead to a diverse work group and solid decision making. Surrounding a transactional leader with transformational leaders, however, will not be as effective. Chief officers, therefore, need to understand the difference between leaders and managers as they promote people. A good manager can make a great training officer or fire inspector requiring transactional leadership, but the same success is not necessarily going to be realized if the manager is placed in a position such as a crew commander, which requires more transformational leadership skills.

Leadership Styles

Leadership is often portrayed as a series of behaviors. Leadership expert Gian Casimir used a historical perspective to define his view of leadership and found that leadership could be classified into the different types of interactions that could occur among group members. They were typed as either task oriented or socioemotionally oriented.

> **Chief Officer Tip**
>
> **Leading versus Managing**
>
> Chief fire officers should not debate whether it is more important to be a leader or a manager. Both skills have their place within the world of fire administration. While chief officers certainly want to become great leaders within their organization, they also must be capable managers of the programs and processes under their control. By taking care of the "business" of management, chief officers can instill a confidence and trust in their troops that directly corresponds to their acceptance of the same leader's ability to lead.

Leadership is a complicated and fluid endeavor filled with subjectivity. A diversely-motivated workforce coupled with varied personalities and styles of leadership can create misunderstandings, rumors, and other misconceptions. So how do we, as those charged to lead our agencies, overcome this? To start, when you take the oath as a chief, it doesn't make you a leader. It places you in the formal position to lead and affords opportunity to do so, but in no way guarantees others will see you as a leader. Depending on your scope of practice, it also allows your department, division, battalion, or station to see all that you do and offer opinions based on their experience, not yours. Opportunities for success in this job are endless, and it's our job to pass on as much as possible so our replacements can start strong and not have to learn what we learned the hard way. There will always be on-the-job training that cannot be replaced with anything but the experience, but if our people know what may be coming because we took time to teach them, the end result is far more positive. Here are a few approaches I have found to be absolutely invaluable in my years as a chief fire officer:

When you take the oath as a chief, it doesn't guarantee others will see you as a leader.

Be competent and able to do your job. There is much talking in the fire service. You need to be the one who can also do the walking. Tactical and interpersonal incompetence is unacceptable and sets a poor example. Go to school. Get all the education you can. Teach. If you operate on the principle of "Because I'm the chief," your effective term of service will be exceptionally short and, honestly, if you subscribe to this approach, you may not even be self-aware enough to realize it. We also need to be emotionally and physically fit. These are rough jobs and they take a toll. Set the example and be the person who takes care of himself or herself while also championing the same for personnel. Be able to do the job you expect your personnel to do. Get out there and train with your people. If you can't, prioritize changing the system so you can.

Communicate constantly. As a chief, you will interact with everyone from the newest fire fighter to senior staff in your department and city. While you may have a preferred communication style, remember it's imperative to communicate to your audience in an appropriate manner. It's not their job to bend to your message but rather yours to send the message in a way they are prepared to understand.

Establish trust. It's one of my favorite statements to our personnel that "we can only move at the speed of trust." Work to be the person whom personnel call when they need advice or help. Be a resource. Reach out far past the fire department and be known in all areas of your jurisdiction and region. Solve problems. Love your job and it will give back in ways you never expected.

Be consistent and offer clear expectations. I'll never be able to say that everyone I work with likes me. I have not found many honest chiefs who can. That said, if personnel know what I expect from them and can consistently expect the same performance and actions from me, we together form a strong force able to meet our mission with excellence.

Realize that your job places you at risk. Your job is to take care of your people and to serve the public in the best way possible. Occasionally, there are no rules for the situation you are dealing with in the station or on the street. It's not often that you can go wrong "doing the right thing." That said, also realize that occasionally you will have to "do what's right." These two approaches can be very different, as the latter can be against policy, contract, or may be the first time it's ever been handled. Regardless, your people need to know you will take care of them and be as fair as possible at all times. Any less is simply unacceptable.

Thank people. Much like communicating, different people enjoy different methods of positive recognition. Be consistent and offer thanks where it's warranted. It's easy to forget that our

VOICES OF EXPERIENCE

(Continued)

departments accomplish spectacular things on a daily basis. I have seen many chiefs do this as the years tick by, making the statement that "It's their job to do those things." Make a commitment to yourself that you will recognize the good as aggressively as the bad. There is far more of it and our people need to know we see the acts they perform.

Be truthful. You're not going to know all the answers, and you need to know who to call and where to look. It's perfectly acceptable to tell people this and get them what they need soon after. You're also going to hit speed bumps in your career, and your troops will see it. You'll feel it in your gut when you know you were wrong. Own these mistakes and help your personnel understand what happened and what you've done to correct the issue. Help them not do the same. Act with integrity and guard that with every action.

Be humble. This pertains to you only. You need to be the absolute best at your craft while not bragging about it. If you're actually any good, others will notice. Conversely, it's your absolute responsibility to ensure you point out every single positive thing you see from your department and members. Explain their amazing work to your customers. Make the media aware of their acts with releases, social outlets, and open houses. The fire service does things everyday most cannot fathom. Don't keep it a secret. The people we protect need to know what we do for them and will better be able to support us if they can even slightly comprehend our responsibilities.

Care. Show up. Be invested in the activities of your crews and your community. Understand what's important to them and remove barriers so they can realize success. Help them through difficulty and promote their success. Stop at lemonade stands. Go to Eagle Scout ceremonies. Go to meetings. Attend block parties. Recognize citizens for their contributions to your mission. Understand your customers so you can serve them better.

Push the boundaries. If you subscribe to the status quo, please don't become a chief. Your people need someone who changes with the needs of the profession and always looks to see what is ahead. I won't lie to you; this will cause you problems in the short term. Occasionally it means changing traditions. Other times, it may be offering a service in a new manner that goes against a long-standing practice. Regardless of what is needed to further your department so your personnel can be better prepared to accomplish their mission, do it. Go after it. It may be tough. It may actually even be painful. It may take a long time, but I give you my word that if you stay the course and never give up you will look back and see what you were able to help your department accomplish. That is the reward.

I'll leave you with a final thought. The fire service needs exceptional people to lead as we continue our fine tradition of exceptional mission execution. We need people who place the department and their subordinates before themselves. We need competence and integrity. Simply said, we need you. Please go after it. Do the best job you can possibly do. Only you know if you have given your all. Fulfill your job to its fullest and guard the integrity of this profession both on and off duty. Thank you for stepping up.

Matthew N. Haerter
Battalion Chief
City of Kenosha Fire Department
Kenosha, Wisconsin

Chief Officer Tip

Organizational Power

Organizational power can provide the leader with the tools necessary to influence the performance of subordinates. Understanding these organizational powers is important to any leader. Recognized organizational powers include the following:

- Legitimate or formal power
- Reward power
- Expert power
- Referent power
- Coercive power
- Informational power
- Ecological power

Chief Officer Tip

Charismatic Leadership

To utilize charismatic leadership successfully, leaders must first build a strong foundation of trust and confidence among those the leader is trying to influence. Charismatic leadership is an important tool if the leader desires to develop subordinates who seek to be self-motivated for the good of the organization.

A chief officer may demonstrate charismatic leadership while implementing a new program or mission to the department members. In this case, the chief officer must energize the employees and convey a positive vision to them that influences the group to support the new goals and work toward success of the new program.

Performance-oriented behaviors, which are aspects of task-oriented leadership, constitute a sizable proportion of supervisory behavior. To achieve organizational success, leadership training should become a priority within the fire service.

What are some of the different types and styles of leadership that members of the fire service may use effectively? Comparing the many different styles of leadership currently being discussed to the criteria for the fire service environment yields the following leadership styles:

- Charismatic leadership
- Situational leadership
- Contingency leadership
- Citizen leadership
- Servant leadership
- The transformational/transactional leadership continuum

Firefighting operations should be performed by teams of well-trained, highly motivated fire fighters. Each individual's effort forms a part of the team's overall efforts. The following leadership styles should not be considered necessarily good or bad. Rather, they each can be applied to different situations as needed. Regardless of which style of leadership is considered, each approach must have a deep and abiding concern for the individual at its core.

Charismatic Leadership

The basis for charismatic leadership lies in the strength and example of the leader. In defining charismatic leadership, experts David Nadler and Michael Tushman set the three primary criteria as envisioning, energizing, and enabling the success of military troops. They also discuss the following leadership techniques (Nadler and Tushman 1995, 108):

- Articulating a compelling vision
- Setting high expectations
- Demonstrating personal excitement
- Expressing personal confidence and support
- Seeking, finding, and using success
- Empathizing

These competencies provide an understanding of the power charismatic leaders possess. These leaders have a vision of what they wish to accomplish. They exude an enthusiasm that inspires others around them to join in a pursuit of their vision. They are then wise and capable enough to allow their personnel to do the best job possible.

To consider this as a style of leadership consistent with the fire service, it must be established whether this style of leadership has at its center a concern for the task or a concern for the individual. Nadler and Tushman also stress that certain limitations exist with this charismatic approach to leadership and that they may be directly attributable to the failings of the leader, rather than to the failings of the task or the people involved:

1. The leader sets unrealistic expectations.
2. People become dependent on the leader.
3. People become reluctant to disagree with the leader.
4. The leader must deliver an aura of continuing magic.

This approach is limited to the skills of the individual leader. The following attributes of the charismatic leadership school of thought suggest that both the task and the people who will accomplish the task are placed at a level lower than the leader:

- The leader is strong.
- The leader has a vision.
- Followers can believe in the leader.
- The leader energizes the group.

Most of the attributes and criticisms found within this style of leadership have a large impact on the people who must be motivated to accomplish a task.

Situational Leadership

Leadership and management experts Paul Hersey and Ken Blanchard created a document at the United States Military Academy explaining their approach to situational leadership. Situational leadership is an attempt to demonstrate the appropriate relationship between the leader's behavior and a particular aspect of the situation—the readiness level of the followers (Hersey and Blanchard 1995). In this style, a leader's success depends on his or her ability to understand and read the readiness of his or her followers in any given situation, hence the name *situational leadership*.

Chief Officer Tip

Situational Leadership

The value of using situational leadership is that it allows the leader latitude in deciding the type and amount of individual subordinate supervision necessary for any given situation. Leaders understand that they must be prepared to adjust their leadership styles at any time with any given employee.

The chief officer must adjust his or her leadership style to the situation of the individual subordinate being led while also taking into account the task at hand. For example, an employee struggling to perform a task may need a more supportive leader, whereas one performing fire ground tasks may expect to operate under a more controlling, autocratic leader. The delegating leadership style is appropriate when leading employees who have demonstrated the ability to perform at a high level without supervision and are judged to be competent in the task assigned.

To define the style more precisely, they state "in Situational Leadership readiness is defined as the ability and willingness of followers to perform a particular task" (Hersey and Blanchard 1995, 207–208). It is not enough for the leader to know how to deal with people; the leader must also be sure that his or her followers are trained and willing to do the task in question. The situational leadership style therefore requires a great deal of day-to-day interaction between leader and follower to ensure that the follower is ready when the time comes to act.

Selected elements of the situational leadership style could be used in fire service training programs directed at situational environments. The following attributes of situational leadership relate well to leadership in the fire service:

- Focusing on the relationship between leader and follower
- Understanding the role of the follower and the role the leader plays in creating effective followers
- Blending task and follower to create the greatest impact for the situation encountered

Contingency Leadership

Another theory similar to situational leadership is that of contingency leadership, proposed by Fred Fiedler, an expert in industrial and organizational psychology. The basic premise of contingency leadership involves the relationship between task orientation and relationship orientation. In both the situational and contingency theories, the willingness of the follower to perform is an important element. In the contingency approach, however, issues of task orientation are paired with the relationship the follower has with the leader. In situational leadership, the willingness of the follower to perform is based more on the needs of the organization. Both theories speak to the interaction among leaders; followers; and organizational tasks, goals, and objectives.

Chief Officer Tip

Contingency Leadership

A chief officer might use contingency leadership in taking advantage of the natural competitiveness of fire fighters when implementing a new health and wellness program. In this case, the competitive spirit might drive individual fire fighters to improve their own personal physical readiness while the ultimate goal of the program is to improve the overall health and wellness of the organization.

Contingency leadership is critical to functions performed under times of stress because the needs of the moment are not known until the emergency occurs. In Fiedler's contingency theory, leadership effectiveness depends on matching the leader's style of interacting to the interaction and influence of the group (Utecht and Heier 1976). The focus, therefore, is on the people accomplishing a task, rather than on the task itself.

The importance of the leader in creating a healthy, competitive spirit is a key to organizational success. In the contingency leadership style, the leader is responsible for creating the proper situational environment in which people can derive satisfaction from their efforts. This requires the leader to understand the demands of the tasks at hand and to motivate the people to work toward a common goal. The following attributes of contingency leadership demonstrate how to use this style within the fire service, focusing on the impact on people rather than on the task:

- Competition (when appropriate) to strengthen the team
- Knowing the members to create a solid team
- Concern for team members
- A thorough understanding of the organization and its task structure

Contingency leadership can be used to improve the organizational climate in fire departments, as long as the chief officer understands both the merits and any possible problems.

Citizen Leadership

Citizen leadership, too, favors the people over the task, although it is a much different leadership style from the three already discussed. In a free and democratic society, leaders can sometimes be expected to evolve from within a societal context. In defining citizen leadership in *Leadership and Democracy*, Thomas Cronin discusses the warring conflicts of freedom and authority, suggesting "We love to unload our civic responsibilities on our leaders, yet we dislike—intensely dislike—being bossed around" (Cronin 1995, 305). This outlook creates friction within organizations. Citizen leadership, however, is supportive in nature; members are equals who work together for the common good. Leaders must respect, encourage, and support their followers. Without the support of followers, there can be no leaders.

Concern for the follower is the basic component for creating effective citizen leadership. The need for equal treatment differentiates this type of leadership from the others.

You Are the Chief Officer

With a focus on preparing for upcoming promotional opportunities, you are reviewing your professional development plan to update your education and training. As a shift captain, you have been reviewing the department's job descriptions for the administrative and executive chief officer positions. As you take note of the increased requirements for administrative skills, you consider how to align your plan to meet these requirements.

1. How does a managing fire officer find the time to satisfy the experience section of the IAFC *Officer Development Handbook* for the EFOP if there are no opportunities to get that experience within his or her fire department?
2. What are the more effective pathways to meet the education and training requirements for the EFOP? How many credentials do you need and in how many disciplines?
3. How can becoming active in the community prepare you for a position as a chief officer?
4. What should you consider when deciding the direction of your formal educational plans?

Introduction

Individuals who choose a fire service career will ultimately decide how far up the career ladder they desire to climb to meet their own personal definition of success. Not all will aspire to become a chief officer. For those who do, personal and professional development objectives will play an important role in obtaining their career goals. Chief fire officers are usually most successful when they follow a path of continuous improvement, both personally and professionally. This chapter discusses various topics that are helpful in career development, including planning professional development, establishing education and training goals, and developing traits of success, as well as administrative topics such as promotional policies and procedures.

Fire Officer III

Developing Personally and Professionally

Education and training are important for a safe and rewarding fire service career, but there are many other methods by which a person can develop professionally and personally. A key component of career development is adopting the concept of continuous improvement. Continuous improvement is a management concept asserting that if any organization is to become and remain strong and vibrant, it must continuously improve. The same can be said for individuals. Understanding the importance of continuous improvement, the basic strategies for improvement, and—most important—having the will to improve will help individuals become the best they can be as they strive to reach the top of their career.

Consider the following question: Am I better personally and professionally today than I was a year ago? Chances are, if the answer is "Yes," then whether you realize it or not, you are probably a person enjoying the benefits of continuous improvement. The next question may be more difficult: Have I done all I can to improve myself over the past year? Most people will honestly answer: "No, I could have done more." Continuous improvement is not just a concept, it is a process. And, individuals move this process forward at different rates. This rate of movement is affected by many factors, including: family commitments, outside employment activities, monetary limitations, supervisor support, personal planning, and available opportunities. The obvious statement is that people who are able to move the process forward faster are generally the people who are more successful.

Traits of Success

To improve continuously, a person must look for ways to develop himself or herself in several important areas. Personally and professionally, successful people have many common traits, including vision, courage, technical knowledge, organizational skills, and people skills.

The first trait, vision, is the ability to define one's idea of what kind of person he or she wants to be and how being that kind of person manifests itself personally and professionally. Establishing a vision takes some thought with regard to where a person currently is and where a person would like to be in the future. An honest self-evaluation can help answer these questions. Those seeking promotion must have a clear vision of what the future would hold should they be successful in their career pursuit. Assessing personal and professional vision is important as a person moves through life and career because vision is dynamic in nature. As life and career circumstances change, so can an individual's vision. For example, if the individual finds himself or herself suddenly in the middle of a family change (e.g., marriage, children, divorce, illness), then he or she may have to alter or postpone his or her own career plans as the requirements of a new career position may conflict with family requirements at home.

Another aspect of vision is the understanding that a chief officer is *always* being watched. It is very difficult for the chief officer to separate personal and professional life. Successful chief officers understand the necessity to be above reproach. As an example, fire fighters may understand the necessity for zero tolerance when it comes to alcohol use and response. Off duty, they can simply choose not to respond after having a few beers with friends. However, once in a chief officer role, specifically the fire chief, choosing not to respond may not always be an option. Accepting the role of chief officer comes with the understanding that, fair or not, one's personal life will come under greater scrutiny. No one should be asked to give up all personal pursuits, but being able to strike a balance between personal life and professional conduct is important to any leader's success.

The second trait, courage, is a necessity for all leaders. Courage allows leaders to make the right decisions, which are not always easy decisions. Chief officers know that good decision making takes into account what is the best for the organization and is not simply how to make everyone happy. Even small decisions have the possibility to upset someone, but doing what is right is far more important than doing what is easy. The legacy a chief officer leaves is often built on the courage required to make the many tough decisions over the long haul.

The third trait, technical knowledge, is of obvious importance because every position in the fire service requires varying degrees of technical knowledge—from fire fighters using tools and equipment to chief officers planning and preparing budgets. To do a job safely and efficiently, one must have technical knowledge. While one might expect that knowledge of administrative and management practices is a high priority for a chief officer, one cannot overlook the maintenance of technical fire ground knowledge. This technical knowledge is still valuable to the chief officer, who is expected to command safe and efficient incident scenes. The more technical knowledge acquired, the more likely one will function at a high level of job performance.

The next trait, organizational skills, encompasses many different aspects. The successful chief officer is one who is disciplined in task planning and completion. The ability to plan and schedule cannot be overlooked. Organized communication skills are also necessary. From timely response to inquiries from citizens, supervisors, and subordinates, to the maintenance of email and document filings, the ability to stay current, in-touch, and on-time is a skill that every successful officer must acquire. Sometimes overlooked, the organization of time itself can be either a help or hindrance to professional development. One must learn the concepts behind time management as he or she looks to balance all the demands of the job along with personal/family responsibilities. On top of those demands, one will also need to make time for acquiring the professional skills required for the next promotion through participation in education and training opportunities.

The final trait, people skills, can be defined as the skill of communicating ideas, concepts, and directions to others in such a way that a desired action will be taken. People skills are very important because interpersonal interactions are key to good relationships. The fire service is a career with many person-to-person interactions. These interactions can be internal (within the department or municipality) and external (within the community). A unique aspect of these relationships is that emergency services personnel interact with each other and the community in both nonemergency and emergency situations. They are required to interact with individuals at their most difficult time, such as when experiencing the loss of a loved one or loss of property during a fire incident. Having good people skills and knowing how to adjust to any given situation are essential to that communication.

Developing Traits for Success

Many resources can help develop traits for success. These resources include training and education opportunities, self-help and biographical books, trade magazines, conferences and seminars, membership in fire service organizations, and observation of others.

Education and training opportunities are numerous (and are discussed in greater detail later in this chapter). Taking full advantage of these opportunities can be of great value.

A trip to a local bookstore or library provides a wide variety of topics that can be very helpful in personal and professional development. Self-help books offer information and concepts about self-improvement. Biographical books can provide insight into how other successful people have developed their own traits of vision, technical knowledge, organizational skills, and people skills.

Many high-quality trade magazines for fire and emergency medical service (EMS) personnel can be excellent resources for information on a broad range of topics. Reading trade magazines helps one keep current with the latest concerns and issues facing the fire service and provides good information on how problems are being addressed by other fire service leaders and organizations.

Conferences and seminars, both local and national, are great venues for learning and networking with other professionals. Additionally, there is a great variety of national, state, and local fire service organizations in which a person can become a member. On the national level, organizations such as the International Association of Fire Chiefs (IAFC), the National Volunteer Fire Council (NVFC), the International Association of Fire Fighters (IAFF), the International Society of Fire Service Instructors (ISFSI), and the National Fire Protection Association (NFPA) can be of great benefit. There is also a wide variety of state and local fire service organizations. Fire officers participating in national, state, and local organizations can obtain considerable information and develop extraordinary networking opportunities. Membership and active participation in these organizations can provide valuable insights on emerging topics, including pending labor management and health and safety legislation, that may directly affect the fire service organization and/or community. New technologies are the topic of spirited discussion among all groups. This explanatory discussion, which often is not readily available to the public, helps the fire officer stay on the cutting edge of innovations in the fire service. These discussions can also provide valuable information to elected officials and stakeholders, improving the image of the chief officer as one who is knowledgeable of current topics.

Observation of others can be one of the best ways to examine the traits of vision, technical knowledge, organizational skills, and people skills in action. Some people may have all these traits and are good examples; others may not. A person can learn from both. Observing others helps reinforce the good behaviors and warns against the bad behaviors.

Personal and professional development over a fire service career and beyond are not static processes that take place with little or no effort. Personal and professional development is hard work. The absolute foundation of personal and professional development is a person's attitude—the way in which a person perceives the world around him or her.

Keeping a positive attitude is no small task in the face of critical evaluations by the people one leads or works for. The key to a positive attitude is to realize that nothing in life or in a career is perfect. Situations will not always go as planned. Realizing this and developing positive ways to adapt to the ups and downs of life and a career are essential to maintaining a good attitude. Fire service leaders must remember that they are responsible for their own attitude, and while they may want to blame a bad attitude on something or someone else, the reality of the situation is that their attitude is what they make it. A person's reaction to a bad situation is just as important, and in fact sometimes more important, than his or her reaction to a good situation. This is especially true in the fire service, where trust in a leader's ability is tested during stressful emergency operations. Reacting positively while in crisis mode demonstrates excellent leadership traits and builds trust within the ranks. How a person deals with adversity also says a lot about that individual's character. Developing good character traits affects attitude in a positive way and makes it easier to maneuver through life and a career in a successful manner.

> **Chief Officer Tip**
>
> **Modeling the Traits of Others**
>
> Chief officers can benefit from examples of both positive and negative traits while observing other fire service leaders in action. It must be noted, however, that just because a particular style of vision or people skills has worked for one chief officer, it may not work for another. Fire officers must take into account the management culture within which they operate and be sure that the chosen style does not conflict with the views or expectations of their own supervisors, boards, and elected officials. Although conflicts in this area might be rare, it is always best to align your chosen styles with your organization.

Fire Service Credentials

Fire officers seeking professional development should not underestimate the value of fire service credentials. Fire officers at the administrative and executive levels need to seek, obtain, and maintain fire service credentials to validate their education, training, experience, and professional development. There are many opportunities for fire officers to obtain master credentials through state fire officer executive programs, the national Executive Fire Officer Program (EFOP) at the National Fire Academy (NFA), and programs abroad through the Harvard Fire Executive Fellowship program.

State Fire Executive Programs

Many states have developed fire executive programs similar to NFA's EFOP. Because many fire officers from volunteer and combination departments cannot commit to the time required to complete the national program, state programs have been developed to meet the professional development needs of fire officers using a more flexible format. These courses may be delivered over the course of several weekends, allowing fire officers from all phases of the fire service—volunteer, part-time, and career—to participate.

NFA Executive Fire Officer

The NFA EFOP is designed for chief officers and junior officers who meet the key leadership entrance requirement and who have earned at least a bachelor's degree. The NFA key leaders criteria allow department members who are not at the chief officer level of the organization to participate in the EFOP. These individuals have been identified by their respective organizations as being change agents for the fire service by demonstrating leadership and professionalism. The EFOP curriculum is a 4-year program that requires attendance at the NFA to complete coursework in executive development, executive analysis of fire service operations in emergency management, community risk reduction, and executive leadership. Following the course work, the candidate must complete original research and submit a paper detailing the findings of the research and recommendations for each of the four courses attended. Obtaining the NFA EFO designation requires a considerable time commitment by the fire officer candidate. The rewards, however, are many—including affiliation with the National Society of Executive Fire Officers.

Chief Fire Officer Designation

Pursuit of the Center for Public Safety Excellence (CPSE) chief fire officer (CFO) designation is not a destination, but a journey. The journey typically begins at the Fire Officer III level. Officers pursuing CFO designation should reference the IAFC *Officer Development Handbook* and the CPSE's CFO designation guide. The handbook and guide serve as a reference to documenting education, training, and experience. They provide a road map for completing the requirements referenced in NFPA 1021, *Standard for Fire Officer Professional Qualifications*. The CFO designation requires fire officers to document their activity in the community and develop a master plan for professional development.

The CFO designation process requires the candidate to choose path A or B in the guide. Path A is for officers who meet the minimum service requirements for education and experience. Path B is for officers who do not have a minimum of either 10 years of chief officer experience or a master's degree. These candidates must complete 20 competencies documenting knowledge, experience, and education in organizational structure, labor management, communications, emergency response, incident management, and data and records management. These competencies are to be verified by evaluators of the CPSE Commission on Professional Credentialing.

The CFO designation application is reviewed by a CPSE evaluator, and a recommendation for credentialing is made. This is a three-year designation; the applicant must maintain credentialing and reapply every three years. Recently, the CPSE developed the fire officer designation to provide aspiring chief fire officers with a progressive step to credentialing development as a chief fire officer. This application is similar to the CFO designation. The candidate must complete the same organizational, educational, training, and experience documentation, including 12 competencies.

Harvard Fire Executive Fellowship

The Harvard Fire Executive Fellowship offers leadership training at the John F. Kennedy School of Government in Cambridge, Massachusetts. The program provides these candidates an opportunity to explore strategy and political management, public value and policy analysis, and internal capacity. Applicants are senior fire officers who have demonstrated significant accomplishments and who have the potential to impact and initiate change in the fire service. They are chosen through an interview process to attend the 3-week course of study.

Fire Service Organizations

Affiliation with fire service organizations is an integral part of an officer's personal and professional development. Fire service organizations can offer many benefits and rewards and at the same time provide the basis for networking with other fire service professionals. Whether at the local, state, or federal level, fire service organizations can provide a valuable resource for tapping into the knowledge, training, education, and experience of others necessary to develop as a chief fire officer.

Beyond membership in fire service organizations, there are opportunities to get involved with committees and boards within these organizations, which is likely to yield greater personal and professional benefits. A fire officer may consider starting slowly by joining a subcommittee or attending conferences and seminars to get to know other members of the organization and its directors and board members. The fire officer should develop a relationship with peers within these organizations and share contact information, as well as seek advice and information on committees and other opportunities to become involved. If desired, the individual should work his or her way up within the organization by seeking opportunities on larger, more visible committees, such as health and safety, technical advisory, or other current issue-based committees. Once comfortable with the organization and its mission, one may want to seek an elected position within the organization. Individuals should develop alliances with regional leaders who will be more likely to offer support to them and their organizations if they get to know them and, more importantly, trust their views and opinions. An individual's investment of time and talent in these organizations will pay dividends later in his or her career as he or she seeks greater responsibilities.

National Fire Service Organizations

National fire organizations provide fire officers with training, education, and leadership experiences while providing national perspectives that affect the state and local fire service organizations. Affiliation with organizations such as the IAFC, IAFF, ISFSI, International Association of Arson Investigators (IAAI), and NFPA assists fire officers with continuing education opportunities and helps officers stay in touch with best practices and emerging technologies in the fire service. Affiliation with national organizations also allows chief fire officers to develop and maintain networking opportunities with other fire officers nationally and internationally.

State Fire Service Organizations

Chief fire officers should develop relationships with state fire service organizations such as the state fire chiefs association, state fire fighters association, state EMS organizations, and volunteer fire councils. Like national affiliations, these organizations can also provide training, education, and leadership experiences. Participation on the state level allows chief officers to become proficient in state affairs and monitor the potential for fire service legislation and statewide trends to impact their organization.

Local Fire Service Organizations

Local fire service organizations provide the foundation for local networking and can lead to further organizational affiliations at the state and federal level. It is advisable for chief fire officers to join their local or county fire chiefs associations. Regardless of whether these associations are formal or informal, they provide the basis for regionalized operations such as central dispatches, automatic and mutual aid agreements, and uniformity in policy and procedures (including standard operating guidelines). Local involvement also helps develop camaraderie and future relationships for personal development. Organizations at the local level may also include emergency medical associations and union organizations (such as the local IAFF unit), especially those that are affiliated with fire departments. Local organization affiliation also allows chief officers to become active in local issues, pool resources, and provide departments with the opportunity to develop regional educational and training opportunities.

Fire Officer III and IV

Professional Development

Professional development is the key to helping chief officers and their personnel achieve their full potential. Every member of a department wishing to be successful has the need and potential to improve individual knowledge, skills, and abilities. With the many positions available within today's fire service organizations, members may have a specific career goal at different levels within the department at different times during their career. Additionally, leaders know that every member of a department is important to the delivery of a high-quality service to the public. The goal of any good chief officer should be to help personnel meet their professional goals and develop into highly valued professionals within the department. Chief officers can also help set the bar within their organization by demonstrating that they themselves have career goals that can be fulfilled by following their own professional development plan.

A professional development program helps ensure that capable leaders are put in positions of authority and responsibility. Capable leaders are absolutely critical for safe and efficient operations on the emergency scene, and the cultivation of these future leaders is absolutely critical to the survival of the organization.

Developing professionally begins by setting realistic and achievable goals and objectives. Chief officer candidates, as well as all other members of the organization, must develop a career path that complements their individual skill sets and internal drive. These goals and objectives become, in essence, a "scope of work" laid out to assist fire officers in professional development.

It is best to begin by performing a personal audit of current knowledge, skills, and abilities. Soliciting the input of peers and superiors can give the individual the honest feedback needed to be successful. If available, personnel can use their department's performance measurement process to gain insight into their strengths and weaknesses. Next, one must evaluate his or her portfolio. A portfolio contains years of experience (experiential learning), training attended, and education. This may include college, technical school, or seminars one has attended. Looking at project and organization involvement, does the individual belong to local, state, or federal organizations, and does he or she participate in seminars, activities, and meetings? Is the individual current on best practices, and does he or she contribute to the community and community groups such as the Red Cross, Kiwanis, Rotary, church, or other civic-oriented organizations?

Once the individual has performed this audit, he or she is ready to compare accomplishments to national standards and models such as those set forth by NFPA, IAFC, USFA's Fire and Emergency Services Higher Education (FESHE), and Center for Public Safety Excellence (CPSE). Given this information and the IAFC *Officer Development Handbook*, one can develop the scope of work he or she must undertake to create a professional development plan. This plan will identify the experience, training, and education one must accomplish to progress in career development.

Chief Officer Tip

Take Your Time

Do not tackle too many of your goals at once; doing so could leave you overwhelmed, disappointed, and disillusioned. Frustration can cause delays in personal development. It is documented that 33 percent of all EFO candidates drop out before the completion of their fourth year at the National Fire Academy. This statistic is similar to the drop-out rate for many master's degree programs. There are many reasons this happens, including change of employment, change in family life, feelings of being overwhelmed, and burnout. The key is developing a chronological plan that identifies what you want to accomplish, when you want to accomplish it, and how to make it happen.

Successful organizations are ones that are able to provide their personnel with many opportunities for personal development. Those who aspire to advance within the fire service ranks should strategically take advantage of opportunities that provide them with the ability to showcase and hopefully increase their knowledge, skills, and abilities. Assertive fire officers will involve themselves in their organization by volunteering for activities, projects, and processes that coincide with their personal development plan. While it is advisable for individuals not to extend themselves beyond their capabilities, most administrators will want their personnel to be successful. Therefore, in an effort to gain additional experience, it is acceptable to take on assignments that push individuals beyond their comfort level. As an example, one might ask to be part of a committee that is exploring apparatus development, health and safety research, turnout gear selection, and standard operating guideline (SOG) development. Individuals may also want to seek out higher levels of training as an instructor or a fire safety inspector or fire safety educator. These activities can further demonstrate leadership qualities. As one gains additional experience and knowledge in these areas, he or she will become a greater asset to the organization and at the same time demonstrate drive and determination.

Education and training are the foundation of the knowledge, skills, and abilities of all members of the department. New and future threats to our society (both natural and human made); exponential increases in technology; and increasing demands on fire service organizations to provide more, often with less support in the form of funding, education, and training, have become increasingly important considerations for today's fire service leader.

Developing personally and professionally is important in all areas of life. The key to success is developing a working personal development plan that is both chronological and specific to cover a designated period of time. Adopting the continuous improvement concept described in this chapter can help ensure a rewarding and satisfying future while at the same time spreading the personal improvements over a more manageable period of time.

Properly selecting and promoting personnel in a fire department is one of the critical areas in which a department maintains and enhances its ability to fulfill its obligation to the citizens it serves. The directions taken by members of the organization during their internal and external interactions determine the professionalism of the department. Without good leadership across all ranks, the department is handicapped in its ability to maintain the proper motivational environment for its personnel to be successful. The promotion of personnel is a very important function for any fire service organization because the people who ultimately serve in those chief officer positions will have many critical responsibilities. These responsibilities must not be taken lightly because it is the goal of all chief officers to do the following:

1. Ensure the safe and efficient operation of the department at all levels during all operations, especially emergency response, emergency scene operations, and all training activities.
2. Ensure that the individuals in the response companies are at a high level of readiness. This includes good training, teamwork, and attitude.
3. Ensure that the members of the department understand the overall mission of the department, its goals and objectives, the importance of teamwork, and how each individual is a valued part of the department by being the best he or she can be.
4. Lead within the department and create a career environment that helps its members strive to be their very best.

A fire chief once said to a battalion chief, "The speed of the leader is the speed of the shift." If a fire department is to fulfill its mission safely and efficiently, it must promote its best and brightest members.

Chief Officer Tip

Preparing for Promotional Exams

Take the following steps to prepare for promotional exams:

- Stay informed and continue along a path of self-improvement.
- Make sure you understand and comply with all the requirements of the testing process.
- Organize your preparation efforts; do not overwhelm yourself, but rather prepare in a step-by-step process.
- Get help and advice from other chief officers who have been successful in similar processes.
- Review fire service periodicals, textbooks, and legislative materials specific to Fire Officer III and IV that identify and explain best practices in the fire service.
- Use video recording to practice and critique any oral presentations that may be required.
- Prepare topic notes for review prior to any oral interview you may face during the testing process.
- In all of the exercises, give as complete information as possible in the time allotted.
- Remember that preparation is key and should start the day you decide to attempt promotion.

Professional Development Model

The IAFC *Officer Development Handbook* defines professional development as ". . . the planned, progressive life-long process of education, training, self-development and experience"

(IAFC, 1). FIGURE 2-1 shows a nationally recognized model for fire officer professional development. Two critical elements of professional development, education and training, form the basis of the model. From an educational standpoint, the model recommends three levels of degree, one for each of three officer positions: associate's degree for the FO II, bachelor's degree for the FO III, and master's degree for the executive fire officer or FO IV. It is also recommended that the FO III obtain professional designations in preparation for the executive fire officer position: Executive Fire Officer Program (EFOP) and the Chief Fire Officer (CFO) designation.

Fire officers can use the model to map out their own plan for continuous improvement. The model shows growth of the individual through the four phases of officer development: supervising, managing, administrative, and executive fire officer. The model is also supported by the FESHE program and is referenced in the CPSE CFO designation program.

Personal Goals

As children grow, they often dream about what their future will look like. Many of those who chose a career in the fire service developed that particular goal as a child. Perhaps it was a desire influenced by a parent or relative who was also in the fire service, or maybe as a child they experienced a positive and lasting impression from a fire fighter during some time of direct need. Whatever the reason, now that the individual is here, he or she may question what the next step should be. If one aspires to reach the goal of chief officer, then it might be of benefit to review and improve his or her ability to set, and more importantly to reach, personal goals. Developing goal-setting skills and the ability to reach those goals is an important part of career development. Those desiring to someday become chief officers would be wise to learn from those who traveled that path before them.

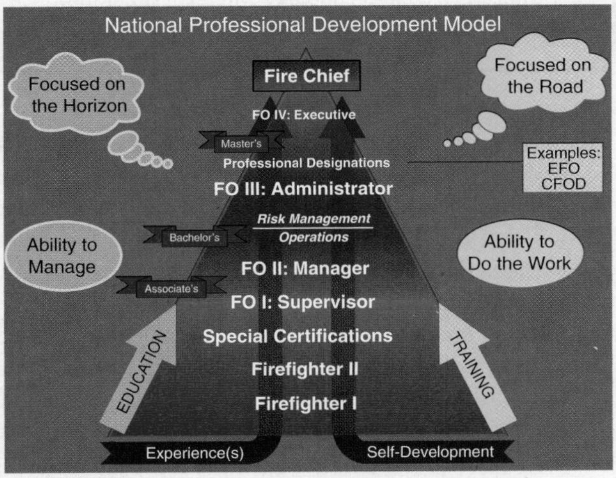

Personal goals can take many forms, including both cognitive and psychomotor skills. Early in our careers we tend to focus more heavily on skill sets that allow us to perform hands-on tasks. Those tasks that would typically be found on an emergency scene and performed by fire fighters under the direction of a company officer are of primary importance. In these early years, basic skill-level trainings are scheduled for us, with objectives and outcomes established by policy and standards and attendance dictated by established training schedules.

A fire fighter who decides that he or she would like to take advantage of additional opportunities for promotion usually is more successful if he or she expands his or her basic skill levels by attending outside trainings that teach advanced tactics or special operations. Thus begins the process of goal setting. For example, a fire fighter preparing for a future lieutenant's exam might be in a better position if he or she has experience in more advanced tactics like hazardous materials or technical rescue and has also prepared himself or herself through advanced education and certification in the area of company officer competencies. While goal setting should begin early on in one's career, it is a skill that, once developed, can serve the individual well into the future as he or she continues to rise through the ranks of the organization.

Goal setting does not end with that first promotion. The process should continue throughout one's career. Goal setting by chief officers also demonstrates to subordinate officers the value of continuous improvement. Chief officers are often looked at to set the bar for their organization; individuals who strive to continuously improve themselves through the goal-setting process can serve to motivate those below them.

For chief officers, goal setting takes on a more cognitive function than is seen in lower ranks. The national professional development model in Figure 2-1 can serve as a guide for setting the goals of an upwardly mobile fire officer. Using this model, fire officers can easily identify both education and training goals for each officer level.

Goals set for chief officers often also involve activities or interactions with people outside the organization. For example, a chief officer might strive to become more involved in his or her greater community. In these cases, involving themselves in activities such as community United Way campaigns, countywide emergency management activities, political lobbying (where and when allowed), and/or membership in various fire and non-fire organizations may provide the chief officer with opportunities to showcase his or her talents, sell the department's service, build networking opportunities, or obtain knowledge not available within his or her own organization. For those who seek to advance, participation in these outside activities may begin with simple membership or committee work but ultimately, through personal goal setting, lead to the

Courtesy of FEMA.

FIGURE 2-1 The National Professional Development Model.

chief officer assuming a leadership position within these outside organizations and committees. For example, a chief officer involved in a United Way campaign one year may seek to obtain appointment to the local United Way board of directors in the future. This would place the chief officer in a position of networking with other community leaders from many different governmental and private agencies and businesses.

When reviewing goal-setting opportunities, chief officers should not limit themselves to fire-related assignments. Successful leaders of any fire service organization know that building a wide knowledge base of outside community functions and adding community leaders in all areas to their contact list allows them the opportunity not only to promote their department to those on the outside but also to put themselves in a stronger political position within their community. This can become especially beneficial when faced with requesting resources, selling their department needs, and making requests to expand services. It is harder to say no to chief officers who are viewed not only as leaders within their department but also within the greater community.

The proper setting of goals can greatly enhance the chief officer's leadership abilities. Setting and obtaining goals provides evidence of a positive work attitude and demonstrates a motivational effort for all to follow. Having goals at all levels keeps the organization fresh and moving in a forward direction. As the old adage states, "If you're not moving forward, you're falling behind."

Chief Officer Tip

Setting Goals Involving Outside Organizations

If a chief officer sets goals relating to involvement in organizations outside of the fire department, he or she should choose his or her areas of involvement wisely. It is easy to overload the chief officer with outside tasks and responsibilities that ultimately take time away from running his or her own organization. This can become a detriment to the chief officer instead of a benefit. It is best to go slow and choose outside involvements that meet these basic requirements:

- Enhance a chief officer's skill set
- Complement services provided by the organization he or she leads
- Accord with existing priorities or goals of his or her supervisors (e.g., elected officials, city manager, township supervisor)
- Provide the chief officer with opportunities to showcase his or her organization
- Serve the greater community good
- Fit into the chief officer's time schedule

Education and Training

The importance of education and training in the fire service cannot be overstated. In the areas of emergency response, the goal of education and training in the fire service is not to create robots that mechanically react to a problem or situation. Safe and efficient emergency response demands sound training programs and educational requirements that help create thinking, feeling human beings who can look at a situation or task, determine the proper course of action, perform that action, evaluate its effectiveness, and adjust to the dynamics of the situation at hand. Personnel must be skilled in the art of situational awareness, strategy formation, tactical implementation of the strategy, and ongoing evaluation. Education and training should help keep chief officers proactive when performing their incident commander role.

The tools and opportunities for education and training in the fire service have certainly expanded over the years. The use of technology and its rapidly advancing ability to provide a variety of learning venues has increased the opportunities to learn. These learning opportunities come in various forms, from online courses to webcasts, from online books and magazines to informational and discussion websites specific to the fire service, from online information searches to remote library access, and so on. With the help of technology, traditional information for education and training has improved dramatically over the years. The use of technology in education and training should not be considered ancillary to the education and training process; instead, it should be viewed as a critical component of the process. Technology has become a vital part of professional development at all positions within the fire service and should be appropriately integrated whenever possible.

Life Safety Initiatives

8. Utilize available technology wherever it can produce higher levels of health and safety.

Education and training in nonemergency areas should have the goal of creating officers with the ability to perform the tasks associated with their position within the department. Most education and training programs on the administration side focus on the performance of tasks that are routinely performed in the department. These may include basic tasks, such as how to review incident reports at the line officer level, to the more complex tasks, such as strategic planning and budgeting, normally at the chief officer level. Of all the topics included in fire service education and training programs, leadership is one area that may not be given the level of consideration and emphasis it deserves. The ability of the officers within the department to provide good leadership is important to the level of success that is reached in all areas of department operation.

VOICES OF EXPERIENCE

The discussion of personal and professional development has been highly debated in the fire service in recent years. The heart of the debate comes down to the importance of higher education among fire fighters. Within the past 10 years we have seen a major increase in fire fighters seeking their Associate's, Bachelor's, and sometimes even Master's Degrees while still working as a fire fighter. There are those who question this "push for papers" as it is called, stating that it doesn't make them a better fire fighter. We have to understand the views of everyone when it comes to this sensitive subject for some.

A framed certificate means nothing if the person whose name is on it doesn't use the knowledge gained in pursuit of that paper.

Do multiple pieces of paper framed on a wall make someone a better fire fighter than the individual who got their degree from the School of Hard Knocks? Absolutely not. This shouldn't even be entertained as a valid question. However, we have to accept that the fire service has changed over the past few decades. The fire fighters who cut their teeth prior to 2000 were brought up in a different fire service than we have today. Development was primarily done by real world experiences of being on working jobs of all types. There are no training aids or classes that can ever replace the education of being inside a working fire and getting the real experience. We have seen the decline of working fires in many areas, however, and those coming up in the fire service now don't get the level of exposure that their more salty counterparts received.

One area wherein we seemed to fail while we got the real world exposures was in preparing fire fighters for promotion by exposing them to the human resources aspects of the position (the administrative stuff no one likes to have to deal with).

In today's world we have unlimited possibilities, it seems, to expose personnel to human resource management through continuing education. We have to try and make up as best as we can for the shortfalls on the tactical side of things.

Every fire fighter should strive for personal development within the profession. This development may lead them on a path to promotion to Company and Chief Officer with higher education, or it may make them the best fire fighter, driver operator, squads man, or rescue technician they can be. A fire fighter who has lost his or her drive to be the best is one who needs to reevaluate whether they are in the right profession.

When we put down others who are trying to better themselves we are insulting our entire profession and all those who have served before us. Organizations also have to remember that their goal is to utilize their personnel to their top capabilities.

At the end of the day, a framed certificate means nothing if the person whose name is on it doesn't use the knowledge gained in pursuit of that paper. The knowledge that was gained along the path of the degree is no different than the knowledge gained by the fire fighter from the incident scene. The only difference is that we aren't issued certificates from incident scenes. If you allow yourself to be intimidated by another just because he or she has that piece of paper, then you need to reevaluate your own knowledge, skills, and abilities.

Remember that knowledge is the power and the key to moving forward. Personal and professional development is vital to the fire service continuing to be one of the most prestigious professions out there.

Billy Floyd, Jr.
Division Chief of Fire Training
North Myrtle Beach Department of Public Safety
North Myrtle Beach, South Carolina

Good leadership helps create a good working environment, which in turn serves to positively motivate personnel to do their best. Good leadership is directly related to developing a department culture that fosters the concepts of professional ethics, continuous improvement, safety, and efficiency. See the "Leadership" chapter for a more detailed discussion of leadership in the fire service.

Education and training programs in emergency response should be developed with the goal of creating personnel who are able to apply the appropriate strategies and tactics to a dynamic emergency situation that takes into account three major considerations:

1. Risk versus benefit
2. Safety and rules of engagement
3. Tactical efficiency

These three considerations are vital to the safe outcome of any dynamic emergency situation. The fire service responds to a myriad of emergency situations that require incident managers to constantly weigh the risk of deploying personnel to a situation that can deteriorate at any moment. The benefit of this deployment must be clearly defined with minimal risk to personnel. When the risk becomes too great for the perceived benefit, fire officers must reevaluate quickly because the rules of engagement have changed. When there is nothing to be gained by an aggressive strategy, we must change that strategy and implement new tactical efforts to support the revised strategy. Tactical efficiency is the key to strategic success. If we have inefficient and ineffective tactical effort, the strategy deteriorates to the point of failure, thus requiring a change in strategy and quite possibly a higher risk to personnel operating at the scene of the emergency. Education and training should be ongoing, particularly in emergency response, to maintain a high readiness level to perform safely and efficiently. Chief officers must be masters of the risk-versus-benefit analysis and be prepared to change strategy as the rules of engagement change.

As important as education and training are to emergency response, it is just as important to the chief officer in carrying out his or her administrative responsibilities. In the case of department administration, chief officers will need to rely on their education and training to survive the administrative and political gauntlet of leading a department. Chief officers, in fact, apply some of the same considerations discussed earlier for emergency response, risk versus benefit, rules of engagement, and tactical efficiency and apply them to their administrative world.

- *Risk versus benefit*: Chief officers need to assess the risks being taken when presenting the potential benefits of new programs, ideas, and initiatives to both department personnel and political leaders. Instead of affecting life safety, these risks can impact job security or a leader's image. Savvy leaders understand the importance of the timing and method of such presentations. Education and training can assist the chief officer in developing the skill sets and tools needed to make presentations more effective.
- *Rules of engagement*: Administrative education and training can give the chief officer insight into understanding the proper process for presenting and implementing change. There is always an order of notification of proposed changes. Respecting the administrative chain of command, a rule of engagement helps keep chief officers out of the uncomfortable position of having their city manager or township supervisor upset when they learn that the chief's new idea was first presented to the mayor or other board members. Maybe the whole board is upset because the chief began soliciting community support without its approval.
- *Tactical efficiency*: Chief officers should be able to utilize their education and training to become more efficient in carrying out the tactics of administration. Every good leader knows the value that comes from the efficient use of his or her time. Being buried in tasks usually means missing appointments, assignments, and deadlines while increasing stress levels. Becoming efficient means the chief officer is taking advantage of the skills taught in administrative education and training programs including the use of delegation, time management practices, and basic scheduling and organizational skills.

Establishing Education and Training Goals

It is necessary to establish and evaluate education and training goals so that department personnel can achieve and maintain the required proficiencies. Employees are entitled to a notice of expectations for the various positions in the department. Adequate notice allows employees to develop their own professional development plan that coincides with the job requirements of the organization.

In establishing education and training goals, it is first helpful to distinguish between the terms *education* and *training*. These words are often used interchangeably but have different meanings in terms of professional development. Thinking in terms of the domains of learning can be helpful in defining what constitutes education or training. The three domains of learning—cognitive, affective, and psychomotor—each have specific areas of emphasis in student learning. Cognitive learning emphasizes the intellectual process. Affective learning emphasizes paradigm and attitude. Psychomotor learning emphasizes motor (hands-on) skills. Although all three domains of learning can be used in education and training, the difference lies in how much emphasis a particular domain is given in accomplishing learning.

Using these definitions, education relates best to cognitive or affective learning, and training to psychomotor or hands-on learning FIGURE 2-2.

Understanding the difference between education and training helps in establishing what cognitive and affective educational learning outcomes are necessary and what psychomotor or training skills are necessary for

FIGURE 2-2 **A.** Cognitive and affective learning focus on the intellectual process and attitude. **B.** Psychomotor learning focuses on the physical skills.

proper job performance requirements. With this perspective, it is easier to categorize education goals and training goals.

In establishing education and training goals, the first step is to determine what education and training are needed for a person to properly perform a particular job within the department. Referring to a comprehensive job description provides a good outline for a particular job function. Along with the internal department job description, one can reference other sources to piece together a more complete picture of education and training needs.

One essential source of information is available in the NFPA standards. The NFPA has 16 standards that specifically reference professional qualifications. Two of these standards, NFPA 1001, *Standard for Fire Fighter Professional Qualifications*, and NFPA 1021, *Standard for Fire Officer Professional Qualifications*, are very important references in determining basic job descriptions. There are NFPA standards available for a variety of other specific fire service positions that are considered to be specialized in nature **TABLE 2-1**.

NFPA standards can also help in the design and delivery of training. These national consensus standards are essential considerations in educating and training fire department personnel. In fact, the Fire Officer III level requires that the officer meet the job performance requirements in Chapter 6 of NFPA 1021. The job performance requirements in Chapter 7 of NFPA 1021 must be met for a fire officer to reach the Fire Officer IV level.

Other important sources of information can come from national, state, or local organizations that are involved in education and training delivery and/or certification. Many of these national, state, and local organizations also rely on NFPA standards for guidance in their programs. Today, many training programs and curricula used to educate and train fire service personnel use NFPA standards to develop their textbooks and training materials.

Table 2-1 Specialized Fire Service Positions Represented with NFPA Standards

- Fire fighter
- Fire officer
- Incident management system personnel
- Safety officer
- Hazardous materials
- Fire marshal
- Fire service instructor
- Public fire and life safety educator
- Public safety telecommunicator
- Fire apparatus driver/operator
- Airport fire fighter
- Marine firefighting for land-based fire fighters
- Technical rescue
- Fire inspector and plan examiner
- Fire investigator
- Wildland fire fighter
- Emergency vehicle technician
- Fire brigade member
- Traffic control incident management

Chief Officer Tip

Training Certifications

When choosing a training program to enhance your skills, it may be beneficial to inquire whether the training program is International Fire Service Accreditation Congress (IFSAC) or Pro Board accredited. Obtaining one of these accreditations may allow the training certification to be recognized by another state, allowing the certification to transfer without having to repeat the training.

Educational Degrees

Several levels of college degrees are generally available. The degrees can be categorized into five basic levels based on educational difficulty and curriculum. The degree levels are identified in the list below and include the corresponding officer rank as identified in the National Professional Development Model discussed in the "Leadership" chapter:

- Certificates (e.g., basic fire fighter)
- Associate's degree (managing officer; e.g., lieutenant, non–line officer in supervisory role in department bureau or division)
- Bachelor's degree (administrative officer; e.g., captain or battalion chief, non–line officer in a bureau or division management role)
- Master's degree (executive officer; e.g., all chief officer positions)
- Doctoral degree (not currently identified)

There are also different categories within each degree level based on curriculum and credit hours needed for the degree. For example, degrees awarded in the arts category, such as associate or bachelor of arts, and degrees awarded in the science category, such as associate or bachelor of science, differ in their basic approach to education. Education in arts is more conceptual in nature than is education in science, which is more technical.

Courses are measured in credit hours, which are based on course content and the time a student spends learning the course content. Institutions base the credit hours on a semester system that can be two, three, or four semesters per year. The number of credit hours varies from institution to institution. Approximate credit hours needed for the various degree categories are shown in TABLE 2-2. These hours are cumulative as education credits are awarded. There are three basic venues in which education degrees can be earned, which are:

1. Traditional: Attendance at a specific location and time is required.
2. Online: The entire program is completed via the Internet.
3. Blended: Programs combine classroom attendance and Internet work.

Educational degrees are offered through a variety of institutions, such as local community colleges, private colleges and universities, and state colleges and universities.

Institutional accreditation is a very important consideration when deciding which degree program to enter. All educational institutions are not accredited equally. This is a very important point because one educational institution may not recognize credit awarded by another institution. This can become a big problem if a person has a degree or credit hours from one institution and decides to transfer or further his or her education at another institution. It is therefore important to ensure that the institution one is considering attending has (and will continue to have) compatible accreditation policies.

Table 2-2 Required Credit Hours

Degree Level	Credits Required
Certificate	Usually 20–30
Associate's degree	64–72
Bachelor's degree	120–128
Master's degree	155–168
Doctoral degree	Varies based on curriculum requirements; heavy emphasis on research

The best advice for a person considering entering into formal education is to be well informed before making the time and financial commitment. Educational information can be confusing because of differences in degree requirements, credit hours, accreditation, scheduling, funding, and so on. One important source of information is the student advisor; talking with advisors from several institutions can provide well-rounded information, leading to informed choices.

It is also important for individuals to have meaningful discussions with their supervisors, mentors, and perhaps administrators concerning the direction in which the organization is heading and how their educational plan may fit into the organization's succession plan. Their support and guidance is crucial to overall success. If the plan fits into the overall organizational plan, it will work to the person's advantage. If there is no opportunity or support for advancement, however, one may have to amend the plan or pursue a career path elsewhere.

Another important and very helpful source of information is available from the Fire and Emergency Services Higher Education (FESHE) program established by the U.S. Fire Administration's National Fire Academy. The FESHE mission is to "Establish an organization of post-secondary institutions to promote higher education and to enhance the recognition of the fire and emergency services as a profession to reduce loss of life and property from fire and other hazards." Not all institutions offering fire service degrees follow the FESHE guidelines. This is, however, a good source of information to describe the type of courses that promote its mission.

Many other resources provide good information in training, education, and career planning. These sources include the IAFC, which, through its Officer Development committee, produced the *Officer Development Handbook*. The 55-page document resulted from three years of work and is available at the IAFC website. The document can be used to provide guidance specific to the professional development of a successful fire officer.

Another source of information is the NVFC. Part of its organizational mission is to promote education and training for the volunteer fire service. The need for competent, well-trained chief officers is not limited to career departments. Volunteer officers need to perform some of the same administrative functions as their full-time counterparts. Accordingly, education and training is just as important to officers in this type of department as well. Getting detailed information from several sources can help ensure that the student makes the right decision regarding his or her education. Discerning these goals is important in directing a person's career path. Speaking with a mentor and others who have, or have had, successful careers in the fire service can be very helpful.

Planning education is essential and should be in balance with an individual's short- and long-term career goals. Balance is achieved when individuals match education and training to their current job description while at the same time looking to improve their level of education and training to meet the next professional level they intend to pursue.

Another consideration in formal education planning is determining which major course of study to pursue. A variety of options are specific to (or related to) the fire service. For example, a degree in fire service administration or fire science is obviously fire service related. Other degrees such as public administration, business management, human resource management, leadership, organizational dynamics, and the like can be helpful to a fire service professional. Examining career and educational goals can help determine what major course of study to pursue at the different education levels recommended for the various levels of fire service officer. Formal education alone cannot make people successful, but it can certainly provide them with the tools they need to build a rewarding and successful career.

> **Chief Officer Tip**
>
> **Choosing an Educational Institution**
>
> Remember the following points when choosing an institution for formal education:
>
> - There are many degree offerings; find one that suits your particular professional goals.
> - Many colleges and universities offer traditional, online, and blended programs; seek out programs that suit your timeline for participation and completion.
> - Make sure the institution chosen meets the appropriate accreditation standards for current and future needs.
> - Review all the course descriptions and any other requirements for degree completion.
> - Discuss your education goals and objectives with prospective institution advisors.
> - Seek the advice of fire service professionals who have previously attended the college or university you are considering.
> - Remember that time, effort, and financial commitment to formal education are major considerations; obtain complete information so that you can make a well-informed decision.

Maintaining Proficiencies

Fire Officers III and IV must maintain proficiency in many areas. Many states currently require continuing education in fire suppression, EMS, and fire inspection disciplines, but few require annual continuing education for the four levels of officer development. By referencing the administrative and executive fire officer outlines in the IAFC handbook, fire officers aspiring to levels III and IV can develop a working plan for maintaining proficiency in areas of training, education, and experience.

The training component focuses on technology, information technologies, data management, strategic planning, and leading change within one's organization and the fire service. This can be accomplished by taking 300- and 400-level courses (if at the bachelor's education level) and by taking computer coursework through local colleges. Coursework at the National Fire Academy in Emmitsburg, Maryland, can assist with maintaining proficiency in risk reduction, leadership development, and best practices throughout the country.

The education component of the IAFC handbook identifies college coursework beyond the bachelor's degree level. It focuses on developing skill sets for human resource management, psychology, labor-management relations, labor law, and political and legal foundations required to remain proficient.

The experience component of the IAFC handbook identifies several areas in which the Fire Officer III and IV candidate must become proficient. These areas include financial and program management, community relations, planning, professional memberships and contributions, agency operations, and coaching and counseling. Proficiency in all of these areas can be used for documentation in the CPSE's Commission on Professional Credentialing for chief fire officers. This program enables chief fire officers to attain national credentialing status as a fire officer or chief fire officer. Why is this important? Many communities, fire departments, and fire service organizations are recommending national credentialing for their officers as a companion to the CPSE's fire department accreditation. These two programs provide validation for the chief fire officer and the organization as they collectively strive to demonstrate adherence to national performance measurements. In effect, this validates to their respective communities the ability to perform above the fire service standard.

Community Involvement

It is the responsibility of the Fire Officer III and IV to project a positive image of the department to the community. One way to project this image is by attending and participating in community organizations and events. This level of involvement enables the fire officer to gain a greater understanding of the community needs and to be better prepared to develop strategies for service delivery. The fire officer must also be able to communicate effectively with the community, which requires knowledge of community demographics, social and economic challenges, and how those challenges impact service delivery.

The best way for a fire officer to demonstrate a willingness to get involved is by volunteering on community boards, participating in public speaking engagements, developing a relationship with the media, and attending important community functions.

Volunteer Boards

Fire officers who volunteer their time, talents, and treasures on volunteer boards gain considerable insight into the community while demonstrating their willingness to support all aspects of community life. There are many opportunities for volunteerism; the challenge is finding the right board on which to spend ample time and energy without being spread too thin. Striking a balance among family, work, and volunteer efforts requires attention to time management skills. It is important to start slowly and develop at a pace that complements one's professional and personal development plan. Many local organizations have volunteer boards, such as those for training schools, local chapters of civic and fraternal organizations, YMCAs, hospitals, and emergency services.

Volunteering on boards allows people to interact with important members of their community who may be on other boards and involved in other organizations. These associations may be of assistance in the future as they build a coalition of stakeholders throughout their community. The networking that develops from these relationships provides them with the support base for future programs and service delivery challenges. It is much easier to get the word out if people in the community can place a face with a name and know individuals personally.

Opportunities for Visibility

Opportunities for visibility can be plentiful for fire officers. Whether providing information at the scene of an emergency or providing background on training or educational programs, fire officers must become comfortable with public speaking and working with the media. Being visible is an opportunity for the fire officer to inform, to educate, and to participate in the many events and programs throughout the community, and thereby bring positive attention to the organization. With visibility, however, comes great responsibility. The fire officers' conduct on and off duty becomes the subject of great scrutiny, with the standards of professionalism, integrity, and personal ethics at the forefront.

Public Speaking

Each fire officer, especially at the administrative and executive levels, must become comfortable with public speaking FIGURE 2-3. The art of public speaking should start very early in a fire officer's career. There are many resources available to assist fire officers in developing the skills necessary to be successful. Fire officers who become involved as fire service instructors can gain confidence and improve some of the same presentation skills used in public speaking. Instructors learn to become comfortable in front of fire service students, and that comfort level shows when speaking to external groups.

FIGURE 2-3 A chief officer must be comfortable with public speaking.

College-level speech and communications classes provide theory and opportunity to get up in front of an audience and deliver speeches. Participation in civic organizations often presents public speaking opportunities. Advanced programs, such as Toastmasters, provide excellent learning opportunities to become proficient in public speaking. It is vitally important for a fire officer to deliver messages to the community in a comfortable and professional manner. Public speaking provides individuals with a free opportunity to sell their organization and to educate the community about their service.

Handling Controversy

Every chief officer will at some time be required to present in front of a public body. In addition to developing the skills necessary to make an effective public presentation, the chief officer must also prepare for the possibility that the presentation may be met with resistance or even anger by those in attendance. Rarely is everyone 100 percent on board with every idea presented by department leadership. Chief officers may find themselves confronted by upset elected officials and/or members of the public such as citizens and business owners. In dealing with confrontation, the chief officer must remember to remain professional at all times. It is never a good idea to turn the presentation into a public argument. Remember that chief officers are considered the experts on fire service issues; demonstrating a calm demeanor will only enhance command presence in front of the group. The following simple strategies may assist chief officers who find themselves in a difficult situation:

- Instead of debating a point of contention, the chief officer can offer to assemble additional information on the topic and forward the findings to the questioning party at a later date. Follow-up should be completed as soon as possible after the conclusion of the public meeting in order to demonstrate true concern for the inquiry.
- Offer to meet with the individual(s) at a set date and time to personally discuss the issue further.

If it is known in advance that an individual or group will raise objections to the presentation, the fire officer may wish to do one or more of the following:

- Postpone the presentation to allow time to gather additional information to counter any known objections.
- Be sure that individuals who are supportive of the presentation will be in attendance as well.
- Offer to meet with concerned parties prior to the meeting.

Remember that elected officials are the bosses and ultimately report to the citizenry. Even if a person is right, he or she should never use knowledge to embarrass any party involved in a public meeting. "Respectful, empathetic, polite, and calm" should describe one's actions even when under attack.

Media

In addition to public speaking, communicating with and through the media can be very uncomfortable for the unprepared fire officer at any level. During emergencies, fire officers will often need to get important and timely information to the people they serve and protect. Often the only way to accomplish this is through media outlets. In times of crisis, communities and community politicians look to their chief officers to provide the level of calm needed to reassure members of the community that everything is under control. The keys to successful interactions with the media are education and training, planning, and experience.

There are several educational/training opportunities at the local, state, and federal levels to assist the chief officer in media interaction. Public information officer courses provide basic and advanced techniques for interacting with all types of media, along with ample opportunity for role-play and live skill testing. There are also many opportunities to improve media knowledge through course offerings at the technical and college level. There are courses that focus on the different forms of media, how they work, how they are produced, and what it takes to be a journalist. Why is this necessary for the fire officer? Simply because for the journalist to understand the needs of the fire officer and the department, he or she must also understand the journalist's needs.

Planning for media encounters is critical to a chief officer's success as the department spokesperson, and it begins with the development of a media policy for the organization. The policy should outline who is authorized to serve as the department's public information officer, types of information that may be released with and without authorization, how the release is made, and to whom it is released. Additionally, the timing of the release may be critical: in some cases, internal notifications to supervisors, elected officials, and/or other municipal departments should or must be accomplished prior to a general public release being made.

The development of a media contact list is also part of the planning process. Chief officers should identify their media partners in all forms of media: print, radio, television, and online sources of information. In some cases it is also important to identify media deadlines. If the chief officer is able to accommodate media deadlines with the release of information, it can help build positive media relationships.

Planning ahead for media interactions can help present a positive image for the department and strengthen the media relationship for the chief officer. Chief fire officers should look for opportunities to interact with the media in nonemergency events. Planning can include media notification of department activities such as a fire prevention open house or a smoke detector installation program. Advance notice of these types of activities can help representatives of the media plan for the event and schedule it into their own news programs and releases.

Experience can be a wise teacher. Getting to know one's media outlets and the people with whom one interacts is a vital component to surviving in the media world. Experience gained during smaller incidents can help prepare the fire officer for large-scale events when media pressure and scrutiny become more difficult. For example, a company officer handling a press interview on a minor structure fire is getting practice for a press conference held during a much more visible but less frequent event, such as a train derailment, aircraft incident, or natural disaster. Experience can help build confidence and allow the chief officer to display the proper level of command presence as someone of authority who can be trusted and respected.

Although the survival of administrative and executive fire officers is not necessarily predicated on their ability to become involved in their communities, their involvement can certainly enhance their position, helping them survive the gauntlet of public opinion. To be involved means being seen and heard.

Chief Officer Tip

Public Speaking Success

Successful public speakers are those who work at perfecting their craft. The following suggestions may assist the chief officer in becoming a better public speaker:

- Be prepared. Whenever possible prepare the presentation ahead of time by writing it down.
- Dress the part. Depending on the presentation, one may decide between normal day uniform, class A, jacket and tie, or casual wear.
- Choose the right words. When speaking to non–fire service groups, avoid using technical terms and phrases with which they would likely be unfamiliar.
- For preplanned events, arrive early and well rested, take time to explore the "stage" and sound system, if needed, and any visual aids planned.
- Know the audience. Be aware of and plan for how to handle any hot topics or controversial issues that the audience may bring up.
- And don't forget, it is perfectly acceptable to *practice*!

Chief Officer Tip

Working with the Media

Chief fire officers are well aware of the use and value of a public information officer (PIO) on an incident scene. While the PIO assignment is extremely valuable to an incident commander trying to mitigate an emergency scene, a PIO can also assist the chief officer in promoting the mission of the department. A department PIO can assist in informing the general public of company officers completing officer certification programs, providing the general public with news about a new program or activity being instituted by the department, or introducing a new class of fire fighters to the community. Finding ways to expand the organization's communication with the community it serves should be high on any chief officer's priority list. Designating a department PIO who can work with media outlets to get his or her messages out is a great way of improving external communication efforts.

Chief officers should look for opportunities to be visible. They should become proactive and interact with their media outlets, getting to know them and educating them on their service. Officers must avoid the temptation to be silent and adversarial by providing no-comment responses to questions that make them uncomfortable. Remember that if one elects not to provide the information to the public, someone else will, and in the end a chief officer may look foolish when the media quote him or her saying, "No comment." If one is firm, fair, and factual with the media, they will be apt to be fair with him or her. If the officer does not know the answer, he or she should not guess or fabricate the answer. It is best for the officer to tell the interviewer that he or she will investigate or research the issue and get the interviewer the answer within a particular time frame. Remember that he or she will be expecting a return call, so don't forget to reply with the answer in a timely manner, or at least to keep him or her updated as to the status of the research into his or her question. Additional information on working with the media can be found in the "Communications" chapter.

Design Credits: Flames: © Photos.com; Chapter opener: © Jones & Bartlett Learning. Photographed by Glen E. Ellman. You Are the Chief Officer: © Jones & Bartlett Learning. Photographed by Kimberly Potvin; Voices of Experience: © Glen E. Ellman; Smoke: © Greg Henry/Shutterstock, Inc.; Chief Officer in Action: Courtesy of Scott Dornan, ConocoPhillips Alaska.

You Are the Chief Officer Summary

1. **How does a managing fire officer find the time to satisfy the experience section of the IAFC *Officer Development Handbook* for the EFOP if there are no opportunities to get that experience within his or her fire department?**

   Managing fire officers can gain experiential learning through a combination of higher education, NFA programs, Emergency Management Institute programs, mentoring programs, and ride-along programs. One of the advantages to memberships in local, state, and national organizations is the networking that occurs. This networking allows chief officers to develop professional and personal relationships with chiefs across the state and country. Most chief officers are willing to allow less experienced chiefs to participate in mentoring and rider programs to enable them to get the necessary experiential learning.

2. **What are the more effective pathways to meet the education and training requirements for the EFOP? How many credentials do you need and in how many disciplines?**

   The most effective pathway to meet the educational and training requirements for the EFOP is to follow the IAFC *Officer Development Handbook*. This resource outlines the necessary educational components, which can be obtained through the associate- and bachelor-level curricula at most universities and colleges. The additional training requirements can be obtained through local, state, and national organizations, such as the IAFC, IAFF, NFPA, CPSE, and NFA. The areas of focus should be incident management, human resource development, records management and analysis, emergency operations, community outreach, and legal and legislative processes.

3. **How can becoming active in the community prepare you for a position as a chief officer?**

   Volunteering on community boards and joining local organizations can help prepare you for accepting the role of chief officer. These involvements allow you to interact with community leaders outside of the fire service. They provide you with the opportunity to serve as a department spokesperson, selling your programs and mission and improving your public speaking skills. These types of interactions allow you to increase the size of your network by making valuable contacts with those who may be able to support your programs and vision for your department when needed.

4. **What should you consider when deciding the direction of your formal educational plans?**

   Formal education should be pursued under a well-prepared plan. It is best to explore and evaluate potential institutions before making an informed decision about which you will attend. Consider whether the institution is accredited and whether classes and degrees would transfer to other institutes for higher learning if the need arises. You should carefully review the knowledge, skills, and abilities listed for your intended promotion in order to choose the most appropriate level of degree needed (associate's, bachelor's, or master's), whether an arts or technical degree is best suited, and what major subject area best positions you for the future. Finally, consider how the formal education is delivered (classroom, online, or blended) and weigh that information against any time management concerns you have.

Wrap-Up

Chief Concepts

- Individuals who choose a fire service career will ultimately decide how far up the career ladder they desire to climb to meet their own personal definition of success.
- A key component of developing personally and professionally is adopting the concept of continuous improvement.
- Traits of success include vision, technical knowledge, organizational skills, and people skills.
- Fire officers at the administrative and executive levels need to seek, obtain, and maintain fire service credentials to validate their education, training, experience, and professional development.
- Fire service organizations can offer many benefits and rewards and at the same time provide the basis for networking with other fire service professionals.
- Professional development is the key to helping chief officers and their personnel achieve their full potential.
- Fire officers can use the National Professional Development Model to map out their own plan for continuous improvement.
- While goal setting should begin early on in one's career, it is a skill that, once developed, can serve the individual well into the future as he or she continues to rise through the ranks of the organization.
- Safe and efficient emergency response demands sound training programs and educational requirements that help create thinking, feeling human beings who can look at a situation or task, determine the proper course of action, perform that action, evaluate its effectiveness, and adjust to the dynamics of the situation at hand.
- In establishing education and training goals, it is first helpful to distinguish between the terms *education* and *training*.
- Educational degrees can be categorized into the following five basic levels based on educational difficulty and curriculum:
  - Certificate
  - Associate's degree
  - Bachelor's degree
  - Master's degree
  - Doctoral degree
- Institutional accreditation is a very important consideration when deciding which degree program to enter. All educational institutions are not accredited equally.
- The Fire Officer III and IV must project a positive image of the department to the community, for example, by attending and participating in community organizations and events. This level of involvement enables the fire officer to understand what the community needs and to be better prepared to develop strategies for service delivery.

Hot Terms

Commission on Professional Credentialing A national credentialing program developed and evaluated by the Center for Public Safety Excellence.

Continuous improvement A concept that states that for an organization or individual to excel in any endeavor, that organization or person must continually improve.

Professional development A long-term, multidimensional process that encompasses the traits of success: vision, knowledge, and people skills. Through the development of these traits, fire officers will be better prepared for the changing fire service.

Rules of engagement A set of conditions developed by the International Association of Fire Chiefs (IAFC) to guide incident commanders in the decision-making process at the scene of an emergency.

Walk the Talk

1. You have mastered your craft as a technically capable supervising officer. Although tactically strong, as you look to the future you find yourself lacking in the cognitive areas of the chief officer positions in your department. You decide you need to update your professional development plan, and several questions come to mind: What direction should your formal education take? How can you build a network of professionals both from within the fire service and from your own community? What needs to be considered when setting personal goals?
2. A concern with the CPSE Chief Fire Officer Designation process is that it appears focused on career municipal employees. How can a township volunteer fire chief, who runs a community-based business, construct a plan to demonstrate competency in the 20 proficiency areas?
3. As a chief officer, you are tasked not only with representing your department, but also with promoting it. You must educate your elected officials, sell your programs and mission to the community, and respond to media inquiries. All of these activities can place you at the center of attention in front of an audience that awaits your words of wisdom. What can a chief officer do to improve his or her public speaking ability and thus project a positive and professional image?
4. Your elected officials support the concept of their department heads becoming involved in professional organizations, providing they support the mission of their departments and add to the professional development of the leader. What fire service organizations can chief officers join on a national, state, and/or local level, and how can those organizations assist the chief officer in his or her professional development?
5. Develop a career plan for your department's administrative officer positions that lists your strengths and weaknesses related to your knowledge, skills, and abilities. Upon completion of the plan, discuss it with your supervisor for feedback and/or implementation.

CHIEF OFFICER *in action*

You have decided on a personal goal of obtaining the rank of chief officer within your department. As you consider your next steps in this process, you determine that you need to formalize a professional development plan that will prepare you for future promotional opportunities. In preparing your plan, the following questions need to be answered.

1. What level of degree is recommended by the FESHE National Professional Development Model?
   A. Certificate
   B. Master's
   C. Bachelor's
   D. Doctorate
2. National accreditation for chief fire officers that includes an examination of community service is provided through which organization?
   A. NFA
   B. CPSE
   C. NFPA
   D. NIOSH
3. What educational and training opportunities are available for today's fire officers seeking to develop personally and professionally?
   A. College and university fire science programs
   B. National fire service standard-making organizations
   C. National credentialing organizations
   D. All of the above
4. Your involvement on volunteer boards enables you to gain a greater understanding of the community needs and:
   A. to be better prepared to develop strategies for service delivery.
   B. to research educational opportunities for advanced education.
   C. to help you decide which opportunities exist for post-employment work.
   D. to impress your elected officials.
5. Which of the following is *not* a trait of professional success?
   A. Vision
   B. Technical knowledge
   C. Political acumen
   D. People skills

CHAPTER

3 Communications

Fire Officer III

Knowledge Objectives

After studying this chapter, you should be able to:

- Identify common communication problems and explain how to mitigate them NFPA 6.1.2. (pp 50–51)
- Discuss forms of communication at the department level NFPA 6.1.2. (pp 51–56)
- Describe the considerations when communicating in an electronic world NFPA 6.1.2 NFPA 6.4.4. (pp 56–59)
- Identify and describe the various methods of external communications NFPA 6.1.2. (pp 59, 61–62)
- Explain the research process NFPA 6.1.2 NFPA 6.2.5. (p 63)

Skills Objectives

After studying this chapter, you should be able to:

- Communicate within a department through policies, directives, and standard operating procedures/guidelines NFPA 6.1.2. (pp 51–52)
- Hold an effective department meeting NFPA 6.1.2. (pp 52–53)
- Conduct effective emergency communications NFPA 6.1.2. (p 55)
- Implement electronic forms of communication for both internal and external communications NFPA 6.1.2. (pp 56–59)
- Conduct research effectively NFPA 6.1.2 NFPA 6.2.5. (p 63)

Fire Officer III and IV

Knowledge Objectives

After studying this chapter, you should be able to:

- Discuss forms of communication at the executive level NFPA 6.1.2 NFPA 6.4.1 NFPA 7.2.2. (pp 63–66)

Skills Objectives

After studying this chapter, you should be able to:

- Communicate effectively with governing boards NFPA 6.1.2 NFPA 6.4.1. (pp 64–65)
- Develop a program proposal NFPA 6.1.2 NFPA 7.2.2. (pp 65–66)

As the fire chief, you are responsible for establishing both internal and external communications. A recently conducted department survey of employees and citizens shows that your ideas and vision for the department are not being received by both groups. Employees have complained of not knowing what is going on or being the last to know, and results from your external survey seem to indicate that many within the community simply have no idea what the fire department does other than respond to fires.

At the conclusion of a public meeting held to discuss the results of a recent citizen survey, it has become clear that your internal and external communications are in need of an overhaul. Your supervisor has instructed you to review your communication methods and options and to prepare a report outlining your plan for improvement. As the leader of the organization, you see this as an opportunity to improve upon a system you inherited. You understand that times have changed, that people today look to many more methods for communicating, and that relying simply on the print media for external messages or bulletin board posts for internal communications is not enough.

1. What methods are available today for communicating with the citizens you serve?
2. What are some of the issues that can arise when communicating with the general public that could negatively impact your message?
3. What communication methods could you consider when seeking to increase the level of internal communications with your staff?
4. What are some of the problems that you might encounter with the following methods of communication?
   - Web based
   - Email
   - Printed newsletter
   - Bulletin board post

Introduction

In the fire service, the delivery of effective communications is a fundamental skill that is critical for mission accomplishment and organizational success. As a fire fighter advances through the ranks, his or her need to be a proficient communicator increases, and as a chief officer it becomes an essential leadership skill. Ask anyone considered to be a successful fire service executive to list the skills he or she relies on most to be successful; not only will good communication nearly always make the list, but it will often be found near the top. The importance of good communication is obvious on the emergency scene. Consider, however, that the fire service is also a profession that requires one to rely on developing and maintaining relationships. Becoming not only good followers, but also great leaders and communicators, is essential to those relationships. Whether it is the boss, a peer, a subordinate, an elected official, a resident, a customer, a union representative, a vendor, an association representative, or a special interest group, one must learn how to communicate effectively **FIGURE 3-1**. Therefore, when composing the message, it is important to consider the audience and tailor the message appropriately. Effective communication includes not only speaking and writing skills, but active listening skills as well.

The basic purpose of communication is to transmit the thoughts in one person's mind to the mind of another person with a minimal distortion of context or intent. Communication is a lot more than simply giving and receiving information; for it to be effective, it must also include an understanding of the message received. Problems with the communications process very often translate into errors, conflict, accidents, and missed opportunities; in fact, most problems for managers and leaders can be traced back to a communication process breakdown.

All forms of communication involve the transfer of information from one person to another **FIGURE 3-2**. The more

© Birmingham News/Landov.

FIGURE 3-1 Good communication skills help the chief officer develop and maintain relationships with others.

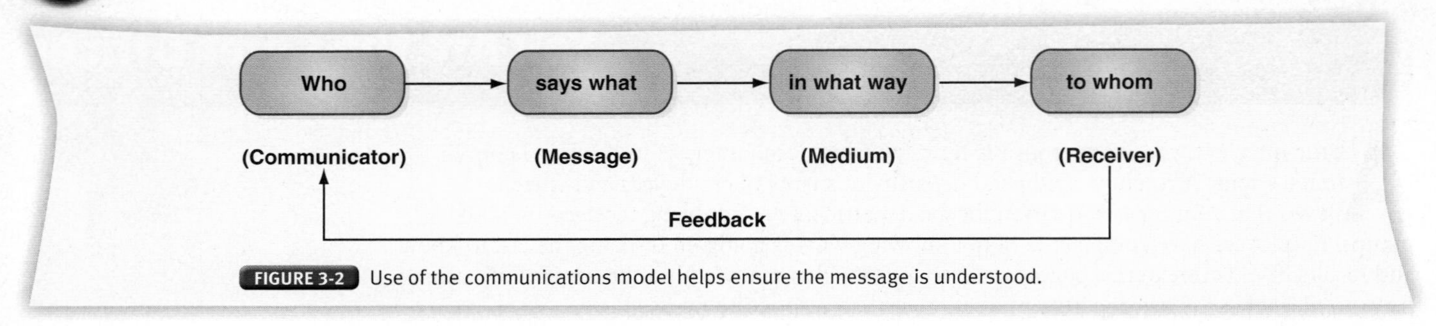

FIGURE 3-2 Use of the communications model helps ensure the message is understood.

correctly the information passes, the better the effect of the communication effort.

It is through the mechanism of feedback that adjustments are made in communications with another person. An emergency service officer must remember that communication is a multidirectional process:

1. *Upward* from the individual member to the upper levels of the organization
2. *Downward* from the center of power to the individual
3. *Horizontally* from peer to peer or division to division
4. *Outwardly* to an audience outside the organization
5. *Inwardly* from an audience outside to the organization

Fire Officer III

Common Communication Problems

Some commonly encountered communications problems are noise, time constraints, filtering, and semantics (meanings).

Noise is an environmental distraction that drowns out or interferes with the message the sender is trying to transmit. Noise can be external to the conversation, such as a radio or TV playing in the background; or internal, created within the mind of the receiver who simply drowns out what the communicator is saying. It can even be as simple as the ringing in a person's ears caused by a hearing defect.

On many occasions, there is simply too little time to get the message across. Or, it might not be the proper time to deliver a particular message. It might also be that the person to whom one wishes to speak is preoccupied with other thoughts or issues.

Filtering is the intentional editing of a message in order to avoid conflict. Many times subordinates filter out bad news when communicating to supervisors to keep their supervisors happy. In other cases, the receiver does not process the bad news being given to him or her because he or she does not want to hear it. Although this is not usually malicious, it can cause some very embarrassing and uncomfortable moments for the person who did not get the complete information.

Semantics deals with word meanings. A communicator should be sure to use language that is familiar to the person receiving the message. Just imagine what comes to the mind of a non–fire service person when hearing phrases like "stretching a hose line" or "catching a hydrant." Even within the fire service, there are regional differences to word meanings. Although the fire service is trying to correct the issue through standardized training, a tanker in one part of the country still means something very different in another. And many fire fighters have no idea what "catching a box" is all about. Always tailor the message to the receiver and be especially cautious when communicating outside the organization.

Frame of reference is another area that can cause miscommunication. Similar to semantics, frame of reference problems occur not because the language is different, but because the meaning associated with the words is different for the sender and receiver. Frame of reference is developed from an individual's background—his or her experiences, perceptions, and assumptions. When the sender and the receiver or an audience do not share a common frame of reference, miscues can easily occur. For example, the chief officer often has a different frame of reference than the council or other governing body he or she serves. Both sides must be careful when crafting communications so that the messages are fully understood and accepted. It is very easy for assumptions to skew the message when there are differences in frame of reference.

Selective listening, or hearing only what one wants to hear, is another common problem, especially in the modern era of information overload and multitasking. Not only does selective listening inhibit the model of effective communications by short-circuiting the sender's message, it often creates hostility on the part of the sender, who is often all too aware of the receiver's selective listening. Trust and respect are necessary elements of good interpersonal dynamics in an organization, but selective listening inhibits the building of trust and respect by breaking down effective communications. Sometimes it is difficult not to have a preconceived notion of how a conversation is going to go; however, it is imperative that one fully listens to the person he or she is speaking with and does not jump to conclusions.

In his book, *7 Habits of Highly Effective People*, author Stephen Covey proposes that we should seek first to understand, then to be understood. A good way to ensure effective listening is to say to the sender, "what I hear you say is . . . ," and then rephrase the sender's message in one's own words. Feedback from the receiver will clarify whether the message was interpreted correctly. Engaged or active, rather than selective, listening on the part of the chief officer will not only ensure effective communication but will also work toward building trust, respect, and effective teams.

Avoid the problem of excessive communication. We are all easily turned off by an overload of information. Our minds can process only so much data at one time. It is critical to remember that listening is the basis for good communications. All of us must work at improving our listening skills to communicate effectively.

Chief Officer Tip

Principles of Effective Communications

Common communication problems can be mitigated using the following principles of effective communications:

- Empathy: Understand the other person's point of view.
- Repetition: Say it enough times to have it understood.
- Trust: Ensure that the source of information is reliable.
- Timing: Do not talk when you should listen.
- Length: Do not use 100 words when 50 will do. Content will be lost in overly long messages.
- Simple language: Take care to make your language easily understood.

Department-Level Communications

Much of the chief officer's time will be spent preparing and delivering department-level communications. Good communication skills at this level are vital to the smooth operation of any organization. Department-level communications can be classified by type, including department policies, directives, and standard operating procedures (SOPs)/standard operating guidelines (SOGs); department meetings; general communications; and emergency communications. Informed employees tend to be better prepared to handle change and deal with the day-to-day issues that arise. The chief officer can head off a multitude of problems by being a good communicator at this level. Employees on the other end of this communications process will feel more involved and will be less likely to rely on information that travels through the grapevine, which can become an unwanted communications process in any department.

Department Policies, Directives, and SOPs/SOGs

While similar in nature, there are also significant differences among policies, directives, and SOPs/SOGs. Policies are generally considered to determine courses of action, whereas directives are regarded as specific orders or instructions that many times have a shelf life shorter than that of a policy. For example, a policy may prohibit certain tattoos or establish how an employee submits a vacation time request. The chief officer might issue a directive to start, stop, suspend, or modify a particular action. For example, a policy might inform department members that only department-issued uniforms may be worn on duty, whereas a directive states that due to an extreme temperature condition, shorts are acceptable for the duration of the heat emergency. It might be easiest to think of directives as the documented form of a verbal order. Violation of either policies or directives carries the force of potential disciplinary action. Often, organizations will develop a set of rules and regulations. This set of rules is often considered policy.

SOPs/SOGs are normally instructive in nature but may not be intended to have the same scope as policies or directives. The key difference evolves from the nature and intention of the document. SOPs/SOGs are designed to provide organizational members with guidance and direction rather than overt management control. SOPs/SOGs are constructed to assist members in making decisions in both emergency and nonemergency situations, but with the recognition that all situations are not "standard" and department members may have to deviate at times from the SOP/SOG. Admittedly, organizations apply this concept in a broad range, with some expecting strict adherence to SOPs/SOGs and others allowing significant interpretation and deviation. Also, while disciplinary action could occur for violating an SOP/SOG—especially when the violation results in some type of loss, damage, or harm—it is not a given that violating an SOP/SOG will result in disciplinary action. For example, many organizations have an SOP covering the placement of response apparatus. Many factors, such as building construction, temporary access blockages, or even traffic, could persuade a company officer or chief officer to deviate from the SOP for good reason.

The policies, directives, and procedures chief officers write may be among some of the most important written communications they will have with their staff. These documents establish the rules, regulations, and expectations of employees in a formal format. Unlike verbal orders, which by their very nature are communicated in person and are open for discussions and clarifications at the time they are issued, written policies, SOPs/SOGs, and directives lend themselves less to the opportunity for discussion because they are usually posted on a bulletin board, found on a web-based information management system, or bound into a manual. Although it is a good practice to review these documents periodically to ensure that employees have a thorough and accurate understanding of them, it is often the case that these documents are visited only when an issue arises and a reference is made to the department's policy, directive, or standard that governs the issue.

One of the greatest challenges with all communications is ensuring that what people mean to communicate is the same as the message the receivers understand. Individuals

Chief Officer Tip

SOP or SOG?

To some, the terms *standard operating procedure* and *standard operating guideline* are considered interchangeable, making it a semantic preference for the organization as to which term it uses. The reality is that more and more often a distinction is being drawn between the two terms. A procedure lends itself to a step-by-step process to complete a task—something to be followed with little flexibility. A guideline can also indicate a path to completion, but implies that the path is more of a suggestion than a mandate, perhaps under "normal" circumstances. In some organizations, because an SOP does not really rise to the traditional definition of *policy* but is still seen by some as written in stone, some have opted, instead, for the term *guideline*, just to avoid confusion.

Officers must understand the ability or authority to deviate from SOPs and SOGs when on-scene conditions require such a deviation. Conditions warranting a change in procedure or guideline are often related to issues of fire fighter safety, tactical advantage, or necessity. When officers deviate from SOPs and SOGs, it is good to discuss reasons for the deviation with staff and management team members so all understand why the change took place and whether changes to the written document are required.

will undoubtedly struggle at first with the task of writing policies, directives, and standards that must be commonly understood. Chief officers should not let this discourage them and should take extra precautions to ensure that they do not become angry or frustrated with their employees if the employees interpret the officers' written word differently from how they had intended.

If you determine that a written policy, procedure, or standard is required, it is a good practice to involve others in its development. The best way to do this is to engage those affected by the problem in a discussion about what should be included in the policy. In some cases, it is best to start with a blank slate, while in others it may begin with a review of an existing policy or examples of similar policies obtained through contact with other similar organizations. Regardless of how a person begins, he or she can ask the members involved for advice on what to include in the overall scope of the document and how to write it so that the end product is understandable to all employee groups. Although under some circumstances this may prove challenging, the individual will often be better off developing a policy that involves the people who will have to live by it rather than writing a policy in isolation and posting it without any input. Although employees may not like the idea of having a new policy, they are likely to be more supportive of a policy if employees are allowed input into the policy during development.

Unless the policy is something that requires immediate implementation (i.e., a policy that must be implemented immediately to ensure fire fighter safety), put the draft of the policy out for the members to review and provide feedback. The chief officer must decide the range of distribution when putting a draft policy out for review. The choices in this case may include any or all of the following: officers only, entire department, chosen representatives of the union, or fire fighter groups. If one puts a policy out in draft form for review and input, it is important to listen attentively to the ideas, feedback, and criticism one receives. Even if one thinks he or she is an excellent communicator, the proof is in whether the employees can clearly understand the purpose and goals of the document. Maintain an open mind to their input and remain flexible to changing the wording of the policy so that the intended message is the received message. It may be dangerous to assume that, regardless of how simply and eloquently one has written the document, it will be commonly understood. Linguistics (the study of the origins and meanings of language) can be complex. Words such as *may, shall, will, can, always*, and *never* may serve to convolute the policies, procedures, and directives. Remember that *always* means *always* and *never* means *never*. Chief officers should be prepared to say what they mean, and mean what they say.

Keep in mind that policies may impact or conflict with the policies of the municipal government under which the department operates. Policies affecting human resource issues may need to be reviewed by city or township legal staff or human resource directors to ensure that they do not run contrary to the municipality's general personnel policy or ordinance. In some cases, a chief officer may be prevented from developing his or her own department policy due to the municipality's efforts to control an issue consistently across all of its departments and divisions. For example, the municipality may decide it is best to have one policy governing the use of social media in the workplace. In any event, chief officers would be advised to understand these issues before presenting a department policy internally and later retracting it due to an overriding municipal concern.

Department Meetings

Many people dread meetings, usually due to past experiences with failed meetings. Properly managed, however, meetings provide an excellent opportunity for improving communications among work groups. The reasons to hold a meeting can vary greatly depending on the objective of the meeting. Officer and staff meetings are normally held to share new information, make announcements, develop or review policy, review incidents, and plan for upcoming events. Committee meetings normally focus on solving a specific issue, but can also include elements of planning and information sharing. Committee meetings benefit from clearly defined roles and role sharing among the group members to facilitate the process. Committees can be formed to tackle short-term issues like turnout gear replacement or ongoing issues like fire fighter health and safety. One-on-one

Chief Officer Tip

When to Write

Do not write policies, procedures, or standards when you are tired, frustrated, or angry. Making the mistake of writing at these times is easy to do. When someone makes a mistake, you may feel compelled to write a policy or directive to ensure that the problem never happens again. Resist the temptation. Instead, examine the guidelines you already have in place that may address the issue. Perhaps you have rules in place that your employee has forgotten about or had misinterpreted. Maybe an incident is an isolated case involving one person and the problem can be addressed with teaching and coaching the individual versus the creation of a formal policy/directive that may affect the entire organization needlessly. So, when something goes wrong, take a deep breath; seek to understand the who, what, when, where, and why of the incident; and then ask yourself whether you really need a policy to correct the problem.

or small-group meetings are often utilized to address specific issues, such as performance improvement counseling. Some rules for conducting effective meetings include the following:

- Have an agenda and make it available in advance.
- Ensure the right people are in attendance. (This is also applicable for webcasts.)
- Define applicable roles such as facilitator, recorder, and timekeeper.
- Be on time, start on time, and end on time; get consensus on any changes to the schedule.
- Understand and work to thwart groupthink.
- Clearly refine any decisions or results.
- Assign tasks and timelines.
- Acknowledge the effort and ensure the participation of all members.
- Follow up on open items.

Agenda Preparation

The first rule from the above list is to have an agenda and to make it available to the attendees before the meeting. It is also good to remember that the person who controls the agenda also controls the meeting. Committee chairpersons or chief officers holding the meeting should accept meeting agenda preparation as their responsibility. The subtleties of the agenda include the fact that where an item appears on the agenda or how much time is devoted to it can greatly influence how the committee views the issue or how much serious discussion is devoted to it. The preparation of an agenda is an important task that must be taken into consideration and handled with care. In fact, if there is no agenda, there may be little purpose in having a meeting. Meetings are often necessary, and an effective agenda is a good way to control the flow of the meeting and ensure its success. Start by clarifying the purpose of the meeting and determining what the outcome of the meeting should be or what the meeting should produce. Not all meetings will have a definitive product. Some meetings are necessary to distribute and discuss new information; however, this is not a license to freelance in the meeting room and skip the agenda and the planning. Below is a guide that will help with organizing an agenda:

- Title: Group name and the date, time, and location of the meeting
- Minutes of previous meetings
- Old business and/or status of pending or action items
- New business
- Date of next meeting

Although not every item on the list will necessarily be included for every meeting type, the list does provide a good framework from which to start. Beginning the meeting planning process early will help the chief officer obtain the best possible results from the meeting. One mistake many meeting planners make is to list every item of old business on the agenda. Think of the old business in terms of what still needs to be addressed; add it to the agenda only if it is still relevant. View the old business as if it were a pending or action-items list. If the item has been completed, list it as such, but time does not need to be devoted to it in the meeting. A table or list in the agenda can cover this and cut down on the amount of time devoted to old business.

For new business, each item should be bulleted out. If necessary, an amount of time may be assigned for action/discussion to enforce time constraints or to control the meeting if confrontation is expected. This step will help keep the focus on moving the meeting forward and will allow the chairperson to move off an issue gracefully if time is needed for other issues. Remember also to publish the agenda in advance and allow meeting members to suggest items for the agenda. In this way, everyone has the opportunity to know what will be discussed at the meeting and can come prepared to present their best ideas. Another important rule is that if it is not on the agenda, it does not get discussed. Except for emergencies, and then only by consensus or approval by the chair, nothing should be added to the agenda after it has been published. Finally, leave some time at the end of the meeting to evaluate the meeting process and get feedback on ways to improve or work on items for future meetings.

A record of the meeting is important to ensure compliance with decisions made, document details of project assignments made, and/or provide a record of information given. The chairperson should appoint someone to serve as recorder of the meeting so that accurate minutes can be taken. The chairperson can then assist with the meeting minute process by making sure that all items discussed are summarized before moving to the next item. The recorder of the meeting should feel free to ask for clarification whenever he or she is unclear about what has been discussed or decided. At the conclusion of the meeting, the draft minutes should be finalized and published as soon as possible so that participants can review them with a fresh recollection of what took place.

Avoid using acronyms and terminology that are familiar only to emergency services personnel. Offer explanations or definitions for all the terms that may be confusing to a layperson. A reader may not clearly understand a concept even as seemingly simple as *total response time*. Explanations may seem basic but may go a long way toward helping others understand the meaning behind the budget numbers. When communicating in writing, it is best to fully explain every detail so that one is able to answer the readers' questions, within the document, before they are asked.

Program Proposals

Regardless of whether a new program is being proposed or an existing one reexamined, the audience of a program proposal can vary depending on the nature of the program **FIGURE 3-8**. Think of a proposal as an opportunity to take an advocacy position for something that will benefit the community or department. However, regardless of how good the idea is or how strongly one believes in his or her program, expect there to be those who are not as excited about the program proposal or individuals who are not as knowledgeable as the presenter on the topic. A program proposal should accomplish the following three objectives:

1. Elicit excitement from the program proponents.
2. Educate the uninformed.
3. Neutralize the objections of program opponents.

The length of a program proposal depends on the complexity of the program being advocated. In writing the program proposal, a chief officer should be succinct and remember to write the proposal for the appropriate audience. Avoid using technical terms, acronyms, and jargon that may confuse the readers, unless the audience is familiar with the jargon. For example, instead of just using the term *SCBA* (self-contained breathing apparatus) or *BA* (breathing apparatus), one may want to refer to this equipment as *the air tanks that fire fighters wear into a fire*. It would be a mistake to assume the terms used by fire professionals on a daily basis are commonly understood by the lay population.

A good format for a program proposal is the three-step process often used by professional speakers: introduction, body, and conclusion. In other words, tell what you're going to tell them, tell them, and then tell them what you told them.

Proposal Introduction

The first step of proposing a program is to tell the reader/listener, in summary fashion, about what the program proposal contains. This is also where the author articulates the problem that needs to be solved or the opportunity and how the community/department would benefit. This is the introduction or executive summary. It contains the highlights of the complete program proposal. The reader should be able to read the introduction/executive summary and get the gist of the entire document. This is important because it is not uncommon for busy people to read only the beginning of a proposal and form their position without ever getting beyond page 1. The author of a quality program proposal may spend days, even weeks, preparing an articulate proposal. After all that effort, the reader should give the program proposal the benefit of a full and thorough read, but remember, in many cases, those reading the proposal may be balancing many competing demands for their time and may have limited time to digest the details and complexity of a comprehensive program proposal. Realizing this can help one carefully craft the executive summary so that the reader can glean the most important points in a short period of time.

The executive summary is used to get the most powerful and convincing points in front of readers right away. This is the reason why newspapers put the most compelling news above the fold of the front page and the television news pitches the most important story at the front of the newscast. Busy people often do not have the time to read an entire newspaper or sit through an entire newscast. Officers should not let this fact discourage them. Rather, they should use it to help them design a strong introduction that can grab the reader's attention and hold it.

> **Chief Officer Tip**
>
> **Response Time**
>
> If you asked people in a room for a definition of total response time, what would their responses be? NFPA 1710, *Standard for the Organization and Deployment of Fire Suppression Operations, Emergency Medical Operations, and Special Operations to the Public by Career Fire Departments*, and NFPA 1720, *Standard for the Organization and Deployment of Fire Suppression Operations, Emergency Medical Operations, and Special Operations to the Public by Volunteer Fire Departments*, define total response time as "The time interval from the receipt of the alarm at the primary Public Safety Answering Point (PSAP) to when the first emergency response unit is initiating action or intervening to control the incident" (3.3.53.6* Total Response Time). You are on a more solid footing when proposing an idea or calling attention to a problem when you can cite national, state, or local standards to back up your position.

FIGURE 3-8 The audience of a program proposal can vary depending on the nature of the program.

Chief Officer Tip

Executive Summary

Use the executive summary of your program proposal to your advantage. List all the positive and supportive points in the executive summary. If there's a downside to the proposal, include it in the body of the report; it is not necessary to make it the front-page news. This way, if readers give your program proposal a brief reading of the executive summary, they are receiving the powerful, positive message you want them to have to support the proposal.

Proposal Body

The second component of the proposal is the body. In this section, the author shares the details that paint the complete picture of the program proposal. Because this section can be long and sometimes complex, special attention should be given to how it is laid out so that it carries the right message (makes the right argument) in a way that makes sense to the reader. Although it may be intriguing and suspenseful in a movie to have flashbacks and flash forwards, such movement in a program proposal can be confusing, if not frustrating, to the reader. A good way to avoid this is to make an outline of how the program proposal should flow. Carefully choose the headings and subheadings so that they flow smoothly. Remember that the author knows where this story is headed and the reader does not. The author must present the material in a logical sequence that allows the reader to fully comprehend each part before moving to the next. The executive summary defines the problem or opportunity. The body shares the details about how the problem is to be solved or how the opportunity will benefit the community or department.

The chief officer can supplement a program proposal with charts, graphs, tables, or pictures as long as they support the argument and are not included simply to fill space or to show volumes of available data. Rather, data should be summarized, using the most meaningful and powerful information to support the argument. For example, it may be impressive to have collected response time data for the last 2000 calls to which the department has responded, but do not try to impress the reader by providing the details of each call in the program proposal. Such detail is not necessary. Synthesize the data into meaningful points and state them in summary fashion, such as averages, means, grand totals, and so forth.

When using charts, graphs, tables, or pictures to support a program proposal, each should be labeled and numbered, and then referred to by the number used. Short (less than one page in length) tables or charts can be used in the body of the proposal. Be consistent with how visual aids are laid out and labeled. When designing a chart or table from scratch, it is a good idea for the creator to give it to someone who is not familiar with the program proposal and ask that person to look at it and explain what it means. If the person can do that without struggle, the visual aid is helpful. If the person struggles to understand and explain the message contained in the visual aid, it should be redesigned to meet the objective.

Longer tables and charts may be included in an appendix in the back of the proposal to avoid drawing the attention of the reader toward studying raw data at the risk of missing the bigger message. Some people like numbers, tables, charts, and graphs; others do not. Including charts, graphs, and tables in the appendix (and referring the reader there) allows those who like visual material to view the appendix as needed. Those who prefer not to use such visual aids or who understand the proposal without visual aids will likely skip reading the supporting documentation in the appendix. Finally, for the visual aids to work at all they must be readable. Small fonts or colors that do not copy well or charts and graphs that contain large numbers of data points that get lost in the clutter should be avoided. A reader should not have to use a magnifying glass to interpret the data.

It is common for a program proposal to have a budget implication, either costing money or saving money. Refer to the discussion on budget proposals in the "Budget and Finance Issues" chapter for help with the financial components. Be sure to use numbers and data in a way that lends clear and easily understood support for the program proposal. Avoid overloading the reader with numbers; doing so can become confusing. Double-check the math to ensure that the numbers add up properly. The author's credibility rests with being accurate.

Often, it is wise to balance the pros and cons of a proposal. As much as one has an idea that he or she is trying to sell, it is wise to address the naysayers up front and deal with the downsides of a program proposal proactively (and every proposal has a downside). The author does not have to dwell on the negative aspects of a program proposal, but acknowledging the potential issues up front and offering some insight on how the proposal works through those challenges will lend credibility to the plan and set the minds of the approving body at ease because the author has acknowledged and addressed the obvious downsides.

Proposal Conclusion

The final component of the program proposal is the summary or conclusion. It is common for this section to include some of the same material as the executive summary. The author can benefit from bringing the reader back to the most important points of the proposal—those contained in the executive summary. The summary should conclude with the recommendations for action (e.g., a budget allocation, a council motion for proposal approval, or an outline of a multistep action plan). Sometimes proposals are for information-sharing purposes only and do not conclude with action steps. The purpose of the proposal drives how the document concludes. If action is desired, it cannot be assumed that the decision maker will be able to infer the needed action. The author must be explicit and list the requested action(s).

Flames: © Photos.com; Chapter opener: © Birmingham News/Landov; You Are the Chief Officer: © Jones & Bartlett Learning. Photographed by Kimberly Potvin; Voices of Experience: © Glen E. Ellman; Smoke: © Greg Henry/Shutterstock, Inc.; Chief Officer in Action: Courtesy of Scott Dornan, ConocoPhillips Alaska.

You Are the Chief Officer Summary

1. What methods are available today for communicating with the citizens you serve?

Departments' options for communicating with the members of the community they serve have and will continue to grow and evolve. Where we previously relied on printed media, we are now faced with the ever-changing world of web-based communications. Printed media and television broadcasts are still important, but are limited by column inches in the newspaper and 30-second snippets on air. Today, departments have found that utilizing web-based communication methods can allow them to control the flow and content of communications to a great extent. These communication methods may include electronic chat rooms, department websites, YouTube, Twitter, Nixle, Instagram, and Facebook, among others.

2. What are some of the issues that can arise when communicating with the general public that could negatively impact your message?

Chief officers must understand the audience with whom they are attempting to communicate. Their frame of reference, educational background, and knowledge of operational terminology can all affect their ability to understand the message being delivered. Recent positive or negative press might also impact the citizens' perspective of your department, leading them to either support or not, and believe or not, the message you are delivering. Timing of the message release might also impact citizen understanding or support. If the community is dealing with the effects of a contentious school millage campaign, then it might be considered bad timing to release information about a new fire department program that may increase department expenses.

3. What communication methods could you consider when seeking to increase the level of internal communications with your staff?

Chief officers have a variety of options to choose from when communicating with staff, including bulletin board postings, internal newsletters, web-based information sites, email announcements, direct conversations, and department-wide or group meetings.

4. What are some of the problems that you might encounter with the following methods of communications?

- Web-based: Information needs to be kept up-to-date. Outdated information leads individuals to lose interest and abandon their visits to the site. Site access may also become a concern. If it is a restricted site, the administrator must be sure to keep the membership roster current. As employees are added to or leave the organization, their access rights must be updated. Additionally, employees must understand that they cannot share their access credentials with others. Sometimes the issue simply boils down to time; while some employees are consistent in their review of web-based information, others find it difficult to find time to review online information.
- Email: Email has become one of the most popular methods of communication, but it is not without issues. Many individuals have more than one email account, which means that senders must be certain that they are sending messages to the right address. Email may also inadvertently end up in the recipient's junk mail folder or be accidentally deleted before being read.
- Printed newsletter: Newsletters, while helpful in distributing information, are usually reserved for less formal releases. Because newsletters are perceived as optional, some employees may never even pick them up. There is also a desire to publish newsletters on a regular schedule, which means that there may be times that information available for inclusion is limited, resulting in publications that hold little interest for the employees. If a regularly scheduled newsletter is cancelled due to a lack of information, and the publication schedule is interrupted, employees may again lose interest or even believe that newsletters are released only when the chief wants to provide his or her view. After missing several issues, a newsletter may be eliminated altogether.
- Bulletin board post: Bulletin board posts are often used to release formal notices. While a popular communication method, they can become lost in a sea of outdated messages if the bulletin board is not kept in an updated and organized manner. You should also remember that while you can post a message, there is no guarantee that employees will read it. Additionally, bulletin board posts may not be good for time-sensitive information. If an employee is off duty for several days or more, he or she would not necessarily have the opportunity to read the posting within a set time constraint.

Wrap-Up

Chief Concepts

- In the fire service, the delivery of effective communications is a fundamental skill that is critical for mission accomplishment and organizational success.
- Some commonly encountered communications problems are noise, time constraints, filtering, and semantics (meanings).
- Department-level communications can be classified by type, including department policies, directives, and SOPs/SOGs; department meetings; general communications; and emergency communications.
- Technology and new electronic devices have greatly affected the world of communications, and that impact extends into the fire service.
- Today's fire officer is faced with countless choices when gathering or distributing information; some opportunities are in traditional venues, but most involve new technologies.
- The first step in conducting research is to clearly identify the problem or opportunity to be defended. Without a clearly defined problem statement, the researcher may gather a large volume of data with limited use to the proposal.
- Being an effective communicator requires more than just reading about how to do it. You have to practice the skills and seek feedback from others to gauge how you are doing.

Hot Terms

Active listening Paying attention to the speaker and eliminating all distractions. It can be portrayed through positive body language and eye contact, asking questions to clarify the message received, and demonstrating genuine interest in the message and the sender.

Bluetooth technology A computer chip and associated program that allow information to be transferred among electronic equipment without the use of wires.

Filtering The intentional editing of a message in order to avoid conflict.

Frame of reference The meaning associated with words based on one's experiences, perceptions, and assumptions.

Linguistics The study of the origins and meanings of language.

Noise An environmental distraction that drowns out or interferes with the message the sender is trying to transmit.

Selective listening Hearing only what one wants to hear.

Semantics Differences in word meanings.

Walk the Talk

1. Complete an online search to locate the websites of fire departments of similar size and operation to yours. Determine how these websites compare to those of your department and which components of others you would utilize in building your own site.
2. Design a template and develop a new letter announcing recent events for your department. Upon completion, provide copies to your line officers and ask them to critique it and provide their thoughts on its value to the organization.
3. Obtain a copy of your state's fire service records retention schedule and review your department's compliance with the regulations.
4. How has social media impacted your department's communications?
   a. Externally
   b. Internally
5. Research social media policies from other agencies and develop a policy for your organization.

CHIEF OFFICER *in action*

As you begin a new year, you reflect on the challenges and problems you encountered over the past 12 months. You realize that there were many issues that could have been handled differently or perhaps even avoided if the issues had been communicated more effectively. In some cases, you thought certain information was distributed in a timely and effective manner, but later found that the information was incorrectly distributed around the department through various social media networks and grapevine discussions. This led to personnel hearing or seeing information that was either taken out of context or slanted against your intentions. In one case, this information was shared outside the department, causing an issue with your supervisor who had to deal with public fallout over the issue.

You make the decision to schedule a series of department meetings with your officers and fire fighters to discuss how to improve the department's communications. Your goal is to improve, control, and expand the communications processes within the department.

1. Which form of informal communications does the chief officer need the most help in controlling?
   - **A.** Bulletin board post
   - **B.** Casual conversation
   - **C.** Grapevine
   - **D.** Email
2. Social media can become an effective communication tool if:
   - **A.** It cannot be effective and its use should be prohibited.
   - **B.** sound policies are developed and information to be delivered is carefully selected.
   - **C.** it is only utilized by employees and not as an official department tool.
   - **D.** it is used for external communications only.
3. When using email for department communications, it is best to remember that:
   - **A.** email is an informal communications method and does not need to be saved after reading.
   - **B.** once sent, emails can quickly end up in the public.
   - **C.** there are no issues with using personal email addresses for department communications.
   - **D.** Both A and B are correct.
4. When utilizing a department meeting for communicating with employees, an agenda:
   - **A.** is not necessary.
   - **B.** should be prepared in advance by the chief officer or chairperson conducting the meeting.
   - **C.** cannot be modified by the meeting members after publication.
   - **D.** suggests the flow of the meeting and allows for other items not listed to be discussed.

CHAPTER

4

Legal Issues

Fire Officer III

Knowledge Objectives

After studying this chapter, you should be able to:

- Discuss the legal existence of a fire department. (p 72)
- Explain the history of law in the United States. (pp 73–77)
- Explain federal laws against harassment and discrimination. (pp 77–80)
- Identify federal laws and statutes regarding rights of employers and employees. (pp 80–83)
- Explain the process of disciplinary action. (pp 83–84)
- Explain the elements and significance of negligence. (pp 84, 87)
- List and define the steps of a lawsuit. (pp 88–90)
- Discuss legal trends that impact the fire service. (p 90)

Skills Objectives

After studying this chapter, you should be able to:

- Take disciplinary action against an employee in accordance with relevant laws. (pp 83–84)

Fire Officer III and IV

Knowledge Objectives

After studying this chapter, you should be able to:

- Identify and describe laws regarding human resources NFPA 6.2.2 NFPA 6.2.3 NFPA 6.2.6 NFPA 7.2.1. (pp 91–97)
- Discuss state laws that impact personnel relations NFPA 7.2.1. (pp 97–98)

Skills Objectives

After studying this chapter, you should be able to:

- Implement an accommodation for an employee NFPA 6.2.6. (pp 97–98)

You Are the Chief Officer

Your department is preparing to conduct an examination for the positions of fire fighter, inspector, and dispatcher. As you prepare the advertisement, you are concerned about the department job descriptions. It seems you need to determine whether the essential functions of the positions are current with actual practice. You assemble a committee comprising fire fighters, officers, dispatchers, and inspectors to review and amend the job descriptions.

As part of the process, the committee is to review and revise the application for the positions. The application was last changed 10 years ago. You need to give them direction by answering the following points in a brainstorming session.

1. What are the essential functions of the position of fire prevention inspector without emergency response duties?
2. Your hiring process for the position of fire fighter will include a physical ability test and medical evaluation to determine fitness for duty. What guidance can you follow when developing a physical ability test that realistically assesses the physical requirements of a fire fighter, and what information is available to a physician who is evaluating a potential fire fighter who has an offer of conditional employment?
3. On the application for employment as a fire fighter, what questions should not be included? Why?
4. You receive a request for accommodation by a dispatch applicant who is farsighted and requires significant magnification and light for reading. How do you determine if the request is unduly burdensome?

Introduction

When you respond to a fire, what gives you the right to exceed the posted speed limit or proceed through controlled intersections (with reasonable care) on the way? Once you get to the fire scene, what gives your fire fighters the right to enter the building? If you make a bad call as incident commander, what gives the property owner the right to sue you? If one of your fire fighters suffers retaliation as a result of a complaint about discrimination, what gives that fire fighter the right to challenge that retaliation in court? The answer to all these questions, and many more, lies within the statutes, acts, laws, and ordinances created through legislative actions at the local, state, and federal levels. Chief fire officers must be ever aware of the sometimes complicated legal issues that affect our decisions and actions on a daily basis. They must also recognize that the training that got them to this point in their career may not be sufficient to successfully navigate the world of legal pitfalls without the assistance of experts in the legal field. There is no shame or weakness in a chief officer seeking the advice of an attorney when confronted with a legal question.

Over more than 200 years of American history, law has entwined itself with how we live, where we work, and what we do. The fire service is no exception. Within the last two generations, chief fire officers have witnessed the decline of negligence protection and the rise of individual rights (which has had particular impact on an institution reliant on tradition and historically organized along paramilitary lines). Most recently, the fire service has seen a handful of fire incident commanders charged criminally in connection with the serious injury or death of fire fighters under their command.

Chief officers cannot perform effectively—or credibly—unless they have working knowledge of the basics of American law, how these laws can affect their work, and how legal risks can be managed. The purposes of this chapter are to provide general guidance for chief fire officers to the legal aspects of fire

Chief Officer Tip

One Size Does Not Fit All

Across the United States, there are 50 state legislatures and state supreme courts, the District of Columbia, Congress, 13 federal circuit courts of appeal, and the U.S. Supreme Court. There are also tens of thousands of state political subdivisions—jurisdictions like counties, towns, cities, and special districts. This chapter cannot—and does not—deal with the statutory and/or common-law issues generated by each of these jurisdictions, but instead focuses on national and significant regional legal issues so chief fire officers can gain an understanding of the general legal landscape.

Competent chief fire officers develop and maintain a relationship with the lawyers representing their department. They do not wait for a legal crisis to meet their own lawyers—they get ahead of legal crises through regular conversations with their lawyers, where each can familiarize the other with information on legal issues and trends, court cases, fire department developments, and what-if discussions (called "hypotheticals" by lawyers).

protection and to link those officers to resources that can help expand and update legal knowledge.

This chapter begins with two simple questions: "What gives fire fighters and fire officers the authority to do what they do?" and "From where do fire codes come?" and links these questions to state and federal statutory and common law. With this foundation, the chapter describes the interaction of the fire service with statutory fire service responsibilities, individual rights, criminal law, negligence, and other areas of the law.

This chapter is written primarily for fire service personnel who work for career or volunteer governmental fire departments. Most of the general areas of law discussed in this chapter apply to governmental fire departments and may apply to private, nonprofit corporations.

Please note that this chapter provides general information on legal issues for chief fire officers and is not intended as legal advice. Chief officers seeking legal advice should consult their department or personal attorney.

Fire Officer III

Fire Departments

What gives your fire department the right to exist? The American fire service is governed by state and local statutory and common law. (In brief, statutory law is law adopted by legislative bodies; common law is law established in appeals-level courts and relied on in future cases.)

If a fire department is a governmental fire department (i.e., part of town, city, township, or county government, or a fire district with taxing authority), there are two sources for its authority. First, at some point in time, the state legislature adopted a statute that gave the town, city, county, or district a general legal right to exist and specific authority to provide fire protection. Then the town, city, or county legislature (town or city council, county board) adopted specific local laws that created the department and authorized it to respond to fires.

These local laws are more than just historical documents. If the local law that governs a fire department refers to *fire* response but the department has since added emergency medical services (EMS), and a trauma patient alleges negligent response by the department, the first question the chief officer may confront is whether the department had the legal authority to deliver EMS. A competent chief officer should periodically review the local laws regulating the department to make sure there is a legal foundation for the changing services the department performs.

A fire department might be old enough to have predated the town, city, or county. In that case, the department was likely grandfathered into legal existence via local government action.

Different local and state laws provide for establishing and regulating nongovernmental fire departments—departments that operate as private nonprofit corporations and contract with towns, cities, or counties to deliver services within specific boundaries. State laws define the minimum requirements for a corporate charter for private nonprofit corporations and often include more detailed statutory regulations (state tax status, minimum number of fire fighters, workers' compensation insurance requirements, etc.) for private nonprofit corporation fire departments. Local laws establish the rules that define the private nonprofit corporation's response area and authorize operation of the private nonprofit corporation fire department.

Special federal tax treatment of a private nonprofit corporation fire department requires the approval of the U.S. Internal Revenue Service (IRS), an approval not required for governmental fire departments. Maintenance of that approval requires annual submission of IRS form 990, *Return of Organization Exempt from Income Tax*. (Other federal and state filings may be required as well.) There are also private for-profit corporations that provide emergency services to units of local government. These are subject to the same requirements as other private corporations and are subject to the ordinary law of contracts, as defined and regulated in each state.

The National Labor Relations Act (NLRA) also applies to private nonprofit (and for-profit) corporation fire departments. One example of the impact of NLRA is its emphasis on employee ability to communicate collectively regarding workplace conditions—pay, supervision, etc. Private nonprofit/for-profit corporation fire departments are thus substantially more restricted by NLRA in their ability to regulate social media use by employees than are public-agency fire departments. NLRA does not apply to public agency fire departments (and state and local government generally).

Fire Marks

Dillon's Rule

Another legal principle that is important to understand is Dillon's Rule, which was first announced in *Merrill v. Monticello*, 138 U.S. 673 (1891). The principle is that a unit of local government has only as much authority as the state legislature grants to the local government or that can be implied as a result of the grant of power. In short, a fire department can do only what the state legislature has authorized it to do through the local government. If the statute says "Yes," it means "Yes." If the statute says "No," it means "No." If the statute says nothing, it means "No." Most states apply Dillon's Rule (or a variation of it) to municipalities within each state (National Association of Counties 2004)

Law in America

North American settlers brought their law with them. The original 13 colonies based their bodies of law on English law, although in Louisiana there are traces of the French Napoleonic Code, and California law still incorporates some Spanish land law.

When the Revolutionary War transformed what had been British North American colonies into the United States, all 13 states had long been operating with legal systems that mirrored the British legal system. When the United States' first organizing effort, the Articles of Confederation, was adopted by a Continental Congress in 1781, a relatively powerless federal government was created. To improve the federal government, the states departed from the British model of common-law rights and responsibilities of citizens and the monarchy and adopted a written constitution in 1787.

The U.S. Constitution defined the structure and operation of the federal government, individual rights, and allocation of power between the federal government and the individual states. Most legal scholars mark ratification of the Constitution as the final act of the Revolutionary War.

The draftsmen of the Constitution did their best to draw definitive lines between state and federal authority and law so that conflicts could be avoided and local issues resolved locally, but their work was not perfect. The evolution of American law would be marked by some of those conflicts, one of which—the question of whether a state had the right to repudiate the U.S. Constitution by seceding—imperiled the very existence of the United States.

Today's American law still comprises common (or case) law and statutory law. Federal and state courts develop and explain common law; local councils, state legislatures, and Congress develop and explain statutory law. State constitutions and the U.S. Constitution provide for judicial review of statutory law, and state legislatures and Congress can amend existing laws or enact new ones in response to what the courts decide. In some cases, these amendments or enactments even overrule previous court decisions.

State legislatures and Congress can also delegate limited statutory-law powers to administrative agencies to adopt and enforce standards of conduct and behavior. In a number of states, legislatures have created fire code commissions and granted those commissions the power to adopt and enforce a fire code as if it were a statute. Because the power to exercise statutory authority is so great, delegation of this power is limited. A fire code commission, for example, must often use a lengthy process involving many opportunities for public input before it can amend a state fire code; however, the fire code is then as enforceable as any statute. Decisions by administrative agencies can be challenged in court and, in some cases, must be reviewed and approved by a state legislature.

Common law is created by courts as they resolve disputes; a description of the dispute, the issues surrounding the dispute, and the rationale for resolution of the dispute are embodied in a judge's decision, which is generally known as an opinion or memorandum. The judge's decision is binding within that court's jurisdiction in any future case involving the same (or similar) facts. Such a binding decision is called a precedent.

Not every court (or judge) has the power to issue decisions that have the power of precedent. Trial courts—federal district courts, state district or superior courts, or local courts—make legal decisions and are the point of entry into the legal system. Those court decisions have no precedential (precedent-setting) power and thus are not binding on any other court. Courts with precedential powers are called appellate courts and may also be referred to as courts of record. The U.S. Supreme Court, U.S. Circuit Courts of Appeal, state supreme courts, and, in a few cases, intermediate state appellate courts would be examples of courts of record.

It is important to note that, with the exception of the U.S. Supreme Court, appellate courts in the United States can establish precedent in a limited geographic area—U.S. Circuit Courts in their jurisdictional areas and each state appellate court in its own state. It is therefore possible for common law to differ from federal circuit to circuit and from state to state. A state supreme court decision can be appealed to the U.S. Supreme Court if there appears to be a conflict between the state decision and federal law. In the absence of such a conflict, a state supreme court's decision is final unless a state's legislature decides to weigh in on the issue. When two or more U.S. Circuit Courts of Appeal decisions are at odds with each other, those cases are often appealed to the U.S. Supreme Court for resolution.

Thanks to a case that arose early in American legal history, *Marbury v. Madison*, 5 U.S. 137 (1803), the U.S. Supreme Court established the principle of the Supreme Court as the final arbiter of the constitutionality of a statute. If the U.S. Supreme Court rules a statute unconstitutional, Congress must defer to the Supreme Court and either amend the statute to correct it or accept the unenforceability of the statute. State supreme courts operate under the same principle.

> **Chief Officer Tip**
>
> **Legal Differences between Career and Volunteer Status**
> Although fire fighters themselves may distinguish between career and volunteer status, legislatures and the courts generally limit those distinctions to benefits (pensions, for example), the employee's property right to the job, and labor relations. Courts and legislatures tend to hold career and volunteer fire fighters to the same legal performance standards (occupational safety and health standards and protection of individual rights, for example) and to the same level of responsibility for respect of constitutional rights.

State Statutory Civil Law

There are 52 entities in the United States with their own bodies of statutory law—the United States, each of the 50 states, and the District of Columbia. While each of these entities has its own unique statutes in place, there are areas of civil law that every state has addressed by statute.

Workers' Compensation

Workers' compensation is a no-fault workplace injury and illness insurance system. When an employee suffers a qualifying injury or illness, the employee receives predetermined compensation.

In most states, the employee cannot sue the employer, although many states permit suits when the employer's behavior is considered reckless, willful, wanton, or gross in nature. Because of fire fighter exposure to products of combustion and other hazardous materials over a career, the questions of which illnesses may be considered work-related and which illnesses are presumed to be the result of work as a fire fighter are on the legislative table in many states. For example, some states have passed laws which presume that certain cancers and/or heart and lung issues are a result of an on-the-job exposure and thus eligible for workers' compensation benefits without the employee having to provide evidence of a specific exposure or cause of the disease. Also, where pension benefits and workers' compensation benefits cover the same illness or injury, the pension codes will provide for an offset of pension benefits for workers' compensation benefits received.

Occupational Health and Safety

Most states, even those that rely on the federal Occupational Safety and Health Administration (OSHA) for routine workplace safety inspections and enforcement, regulate at least some elements of workplace health and safety. As the number of consensus safety standards adopted by OSHA has grown (e.g., the National Fire Protection Association's [NFPA's] fire fighter health and safety standards), application of those standards via state safety regulatory authority is a growing issue. The mechanism for enforcement varies from state to state. New York, New Jersey, Connecticut, Illinois, and the Virgin Islands have separate federally approved health and safety programs for public employees. In those states and territory, the state department of labor, with a federal grant, enforces OSHA standards for public employers who are excluded from federal coverage under 29 CFR 1956. In those states and territory, the federal OSHA program is enforced for private employers. In contrast, 21 states and Puerto Rico have an OSHA-approved state plan for private and public employers. The remainder of the states is subject to federal enforcement of OSHA standards for public and private employers.

Codes and Code Enforcement

The code adoption authority, the local government, and the enforcement authority of the inspectors and investigators are dependent on the state legislature and the authority they give to the inspectors and investigators. That authority can be further limited by the unit of local government. An example of such a limit would be a situation in which the state permits investigators to carry firearms, but the local government employer does not permit such weapons to be carried by employees other than sworn police officers.

The legal aspects of codes and code enforcement are discussed more fully in the "Managing the Code Enforcement Process" chapter.

Fire Fighter Training and Certification Standards

Many states specify the minimum requirements for recognition as a career or volunteer fire fighter and often tie benefit availability and state aid to local fire departments to fire fighter training levels. In states that do not address minimum fire fighter standards, local governments identify and impose minimum requirements. The training requirements are usually based on NFPA standards for professional qualification.

Good Samaritan Acts

Starting in the 1960s, states began enacting laws that limited (or barred) liability of "good Samaritans," people who stopped to help in cases of accident or sudden illness. These laws encouraged passers-by to stop and help, and some states even apply the Good Samaritan approach to persons and agencies that help with hazardous materials incidents. Many of these Good Samaritan laws are limited to off-duty medical professionals (generally physicians and nurses) or volunteer public safety workers. In some states, fire fighters stopping to help while off duty outside their own jurisdiction are protected. In order for the immunity provided by the act to be effective, the individual must not exceed his or her training in providing treatment and must not provide treatment in a reckless or willful and wanton manner.

The introduction of naloxone as the treatment of choice for opioid overdose by laypersons as well as medical professionals has heightened interest in Good Samaritan laws. These laws do vary by state and should be researched by chief officers and other responders.

State Criminal Law

Up to this point, the review of state statutory and common law has focused on state civil law. Both the federal government and states have enacted criminal laws as well. The distinction between civil and criminal law was drawn in Great Britain long before the 13 colonies came into existence.

The fundamental element of civil law is that someone who suffers an injury (in the form of bodily harm or loss of money or property) is expected by society to seek compensation (damages) from the responsible person(s) on his or her own. The common-law burden of proof, the extent to which an injured person had to demonstrate the relationship between the injury and the acts (or failures to act) that were the reasonably foreseeable cause(s) of the injury, was proof by a preponderance of the evidence (sometimes stated as "more likely than not"). The question of whether an injured person would be compensated was a private matter. Until later in the evolution of statutory law, government (whether local or national) played no role and held no responsibility in civil cases.

Criminal law, on the other hand, grew out of a societal conviction that there was some conduct that caused harm to an entire community, not just to an individual, and that the community had a responsibility to prevent such conduct in the future by punishing the perpetrator of the conduct. The model for community response to criminal acts was the historical response to such acts by the Romans and Greeks, and as described in the Bible. Consequently, criminal prosecution was reserved for government on behalf of the community and

VOICES OF EXPERIENCE

Ours is a relatively medium-sized metro department. As you can imagine, 400-plus employees can lead to a steady stream of disciplinary challenges. We are an all-hazards department, including the delivery of both prehospital basic and advanced life support.

One of our rules allows employees to "swap" time in order to accommodate their need for additional time off for personal reasons. The rule states that the employee accepting the time swap must assume the "position"—that is, the duties, responsibilities, and seniority—of the person for whom he or she is substituting.

Over two days of testimony, the nuances of the rules and protocols were belabored.

On the day in question, a fire fighter/paramedic was working for a fire fighter and had assumed his role as the water supply technician on the engine company for that shift. At an extremely chaotic EMS scene, the captain (a basic EMT) was directing resources to deal with two critical patients. His engine company was accompanied on scene by one of our advanced life support ambulances with two paramedics. His crew also had two fire fighter/paramedics, including the individual working on the time swap. During the course of allocating personnel to provide patient care, he directed the individual working the time swap to assist another paramedic who also had a critical patient. The individual balked at the order and was subsequently charged with insubordination.

Our department is in an "at-will" state, but our employees have a property right for their employment under the county merit system and are provided due process as a result of that protection. A predisciplinary hearing was conducted and the three-member panel voted to suspend the fire fighter/ paramedic for 48 hours without pay for insubordination. The employee appealed and an appeal hearing was conducted before the personnel board—a five-member civilian panel—with both sides represented by counsel.

The department considered this a clear-cut case of insubordination: an order was given; it was questioned and not complied with. What we discovered during the hearing was that those unfamiliar with paramilitary organizations were not quite as convinced as we were.

The testimony revolved around two department rules that outlined the employee's obligation to the department and the defendant's assertion that state EMS protocols dictated the employee's obligation to the patient. The department's rule states that the employee (accepting the time swap) has the same obligation to the department and duty as the employee for whom he or she is substituting. Another department rule states that a fire fighter has the duty to take direction from a superior officer. The duty to take direction from a superior officer is echoed in the personnel board rules as well.

The defense relied on a state EMS protocol that stated the highest-level EMS provider on the first-arriving advanced life support unit will assume responsibility for directing overall patient care and continue this function unless relieved by the responding jurisdiction personnel.

Over 2 days of testimony, the nuances of the rules and protocols were belabored. We stated that the city paramedic was employed by us, not the state, and that on the day in question he was a fire fighter, not a paramedic. They countered that regardless of the rule, the employee still held a paramedic license and was obligated to the patient. We countered that on the day in question he was a fire fighter assigned to a basic life support engine company and the first-arriving advanced life support unit was a city fire rescue unit with two senior paramedics assigned who assumed care of the most critical patient. His duty at that time was to follow the chain of command and look to his supervisor, the captain, for direction. He was issued an order to assist another paramedic with an additional critical care patient. Their rebuttal was that the more critical of the two patients could have

VOICES OF EXPERIENCE

(Continued)

benefited from an additional paramedic and the outcome might have been different. Furthermore, they said that the captain, as a basic EMT, was not competent to make that decision.

Our summation painted the picture of a chaotic scene with two critical patients, communication system failures, and non–English-speaking family members. The most critical patient had two paramedics and an EMT rendering care. An additional advanced life support ambulance had been requested, along with a district chief and the department's chaplain. The charge was simple in our minds: He was ordered to assist another paramedic with patient care, he balked at the order but eventually complied, and then migrated back to the first patient, leaving the third paramedic to render care to the second critical patient.

The defendant summarized his role as a paramedic and patient advocate. He painted a picture of his military service and how he had been deployed to a war zone. He understood the chain of command but didn't feel the captain was qualified to issue orders regarding patient care.

The board ruled in the employee's favor, lifted the suspension, and awarded the back pay that had been withheld during the suspension.

What we took away from this hearing was that counsel must be able to fully explain the concept of a paramilitary organization and the importance of the chain of command. What we were unable to convey was that it was not a patient care directive, but an allocation of resources. The captain did not tell him how to render care, only to whom to render care. The other issue we were unable to make clear was rules versus protocols, and that our rule defined the employee's role, whereas the protocol merely provided guidelines for determining scene organization.

F.D. Sims, RN, MPA, REMTP, EFO
Deputy Chief
Mobile Fire and Rescue Department
Mobile, Alabama

negligence suits because if there were too many suits, government would simply stop the activity. Among the examples of governmental activities cited by courts and legislatures was fire protection.

Proprietary activities, on the other hand, were undertaken by both government and the private sector. A citizen could find those services in the yellow pages. Government did not need as much protection from negligence suits in connection with proprietary activities; if there were too many suits, it might be a sign that the government should leave the activity to the private sector. Examples of proprietary governmental functions are hospitals, transit services, and airports.

Another means of drawing the sovereign immunity line was on the basis of the nature of the governmental activity. The discretionary acts of local government, those requiring judgment, should be protected. If they were not, government employees would not make decisions in the face of not knowing whether someone would second-guess their decision via a lawsuit. Fire incident commander selection of an offensive firefighting strategy is an example of an act that has been recognized as discretionary. Ministerial acts, acts requiring little or no judgment, deserve less (or no) protection. A faulty per-room electrical fixture count by an electrical inspector, resulting in the withholding of a certificate of occupancy, is an example of a ministerial act for which a government could be held liable.

Negligence Defenses

In addition to varying levels of sovereign immunity, governments also retained the ability to raise standard negligence defenses. In reading the following defenses, keep in mind that each state has its own collection of defenses to negligence suits. A chief fire officer should become familiar with the defenses specific to his or her state.

Duty Doctrine

To prove negligence, an injured person must be able to prove that the person alleged to have caused the injury owed a duty (duty of care) to the injured person. When a fire inspector inspects a restaurant, is his duty to conduct a competent inspection a duty to the restaurant owner or to the public in general? The bodies of negligence law in a number of states answer that the fire inspector's duty of care is to the public at large, not to any individual. The old-time common-law summary of the duty doctrine is, "A duty to all is a duty to no one." Another basis for application of the duty doctrine is that, were a fire inspector to be liable for an inspection error, that inspector becomes the absolute insurer of a building owner's fire code compliance.

The duty doctrine is limited in some states. A statute can define persons to whom a duty of inspection care is owed, and a fire inspector can create a duty by making representations to a building owner, on which the owner relies. In some states, the duty doctrine defense does not exist.

Protection for Government Employees

In some states, as long as a fire officer (or fire fighter) is doing what he or she is authorized to do, that officer has a defense against a negligence action. If a fire department is authorized to deliver EMS and each officer is certified as an emergency medical technician (EMT) and authorized to deliver EMT services, then a fire officer-EMT applying a splint is acting within the scope of employment. But a fire officer who uses an electroshock weapon to control an unruly patient is arguably acting outside the scope of employment.

Another employee defense is based on the level of negligence. In jurisdictions that assess level of negligence, an employee who commits ordinary negligence, what could be described as simple error, will not be held personally liable. However, if the employee negligence is willful, wanton, gross, or malicious (which is measured by the foreseeability of injury or the relative recklessness of the employee behavior), that employee will stand beside the employer in a negligence action, and a jury will decide the level of liability and damages.

The simplest protection provided by a government employer to an employee is an agreement by statute, ordinance, or contract that the employer will provide legal counsel for an employee named in a lawsuit and that the employer will reimburse the employee for any damages levied against the employee in a negligence suit. Such statutes, ordinances, and contracts often exclude employer protection if the employee was acting outside the scope of employment, is found to have been grossly (or wantonly, etc.) negligent, or was found to have been under the influence of alcohol or a controlled substance.

Government employees can also use the same standard negligence defenses available to the employer.

Assumption of Risk

If a reasonable person knows that a situation or event involves risk of injury, but proceeds to place himself or herself in the situation or event anyway, that person is deemed to have assumed the risk of injury. When the TV cameraperson who slides under the barricade tape at a fire scene to get a better shot of the fire is injured and sues the fire department, the chief officer can defend the department with the argument that the cameraperson assumed the risk of his or her injury and was actually failing to comply with an order not to enter.

Assumption of risk can be (but often is not) an absolute bar to plaintiff recovery; however, it can shift at least part of the negligence burden back to the injured person.

Comparative and Contributory Negligence

A person can play a role in his or her own injury. The comparative/contributory negligence defense stands for the proposition that an injured person should not recover damages for his or her own negligence.

In many states, the comparative/contributory negligence defense means that any damage award to the injured person will be reduced in an amount proportional to the extent of the injured person's own responsibility. If a jury finds that the plaintiff was 35 percent responsible for his or her injuries, any damage award would be reduced proportionally.

In a few states, contributory negligence can bar any recovery. The legal theory in these states is that an injured person responsible to any extent for his or her injury should be ineligible for damages.

Lawsuits

The objective of the American legal system is to handle disputes like negligence claims by mediation, arbitration, or other settlement methods short of a lawsuit. If a dispute cannot be settled and a lawsuit is filed, the American legal system continues to encourage efforts toward settlement prior to trial. Additionally, under the American system, each party is responsible for its attorney fees unless the judge is permitted to award attorneys' fees to a party under the provisions of a statute or agreement.

At both federal and state levels, both pretrial and trial segments of a lawsuit are governed by the legal version of a fire department's standard operating procedures. The Federal Rules of Civil Procedure govern suits in the federal court system, and each state has its own rules of civil procedure for local and state court use. Most recently, the Federal Rules of Civil Procedure and state rules were substantially amended to take into account how electronic records are to be maintained, accessed, and used.

The actors in, and phases of, a negligence lawsuit are straightforward, with little variance from jurisdiction to jurisdiction FIGURE 4-2.

Phase 1: Complaint

A lawsuit starts with a complaint—a recitation of the facts of an injurious event or damage by one or more plaintiff (the injured party), connections between the event and the actions (or inactions) of a defendant (the person allegedly causing the injury), and a request for damages. We noted the four essential elements of negligence earlier in this chapter (a duty, breach of the duty, a causal relationship between the breach and an injury, and injury in fact); a plaintiff must address those four elements in the complaint. The complaint does not have to be detailed because some facts are not known; however, it does need to allege facts and not conclusions. The civil procedure statutes of each state set out the requirements.

In some cases, the defendant will file a motion to dismiss for failure to state a cause of action. Such a motion would be filed if it were clear from the complaint that there are no facts on which the defendant could be held liable. It is also filed if the original complaint is full of conclusions instead of facts. If the motion is denied, the case proceeds with an answer (see the next section). If the motion is granted, the plaintiff may receive permission to amend the complaint to correct the errors. Permission is rarely denied. The complaint is also verified by the plaintiff or an official of the plaintiff corporation as to the accuracy of the facts stated in the complaint.

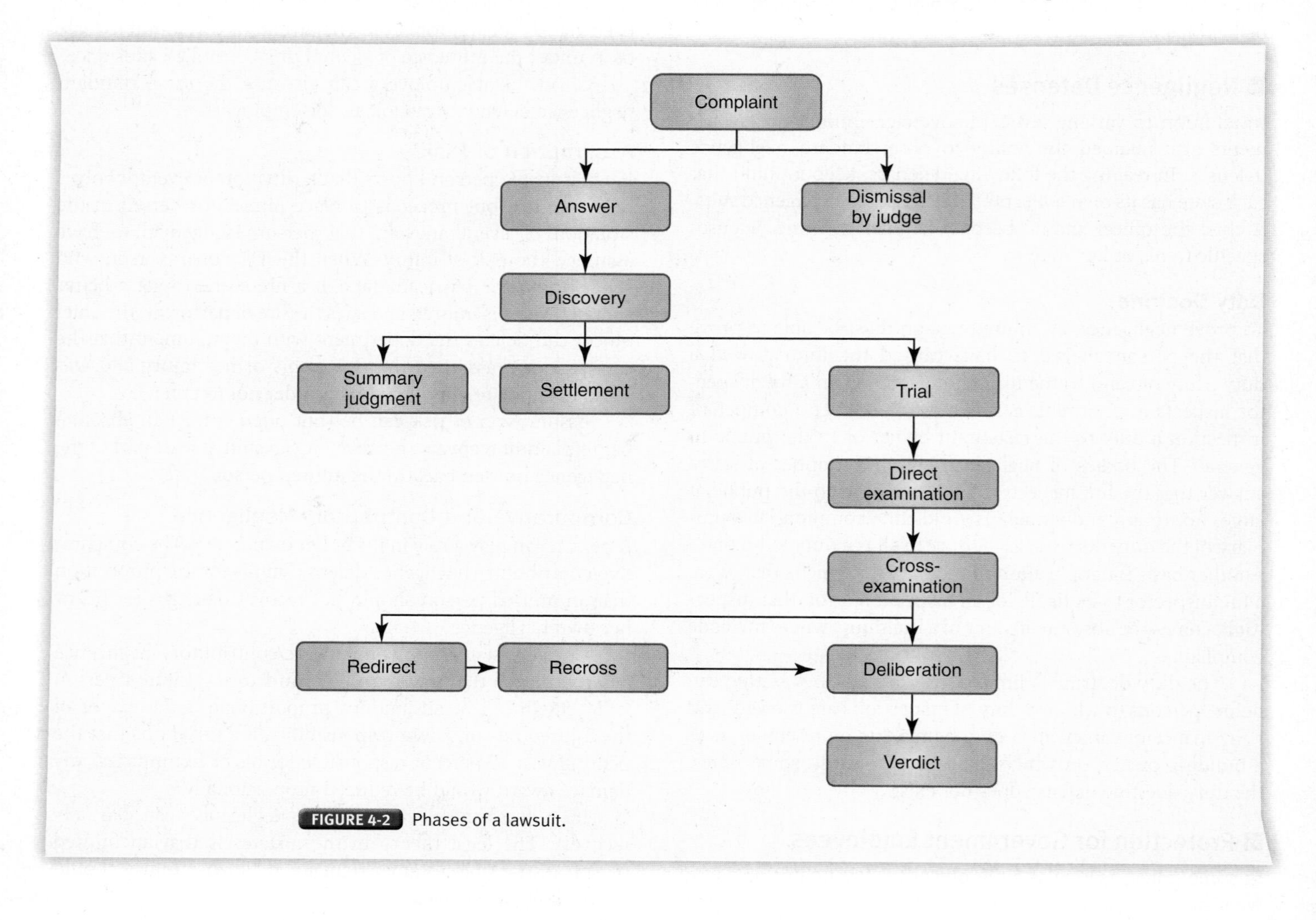

FIGURE 4-2 Phases of a lawsuit.

Phase 2: Answer

The defendant has a specific period of time in which to respond to the complaint. This response is called an answer. Generally, the answer responds to the complaint point by point, noting errors in the complaint, and explaining the event from the defendant's perspective. This is not the time to set out all the facts in favor of the defendant's position. The answers should be concise and not admit anything unless the statement in the complaint is clearly correct, such as "The defendant is the chief of XYZ Fire Department," which is obviously correct. The answer would be: "The defendant admits the allegation in paragraph 1." When all allegations have been answered, the answer is verified by the defendant as to its accuracy.

Phase 3: Discovery

With complaint and answer in hand, the attorneys representing the plaintiff and defendant begin the often long and laborious process of discovery. Pretrial discovery gives each side of a suit access to a wide variety of paper and electronic records and documents held by the other in order to bring the facts of the matter to everyone's attention FIGURE 4-3. Additionally, this process gives each side the ability to discover potential witnesses and requires them to answer questionnaires (written interrogatories) under oath and to allow attorneys to depose (examine and cross-examine) potential witnesses outside a courtroom, under oath.

One of the objectives of the process is that each side will get a look at the strengths and weaknesses of the other and reach a settlement. While settlement is an efficient byproduct of discovery, the primary purpose is to eliminate the "surprise and ambush" tactics that existed before the procedural rules were enacted. This objective requires that once a lawsuit is served on the defendant, all relevant material, on paper or electronic, is protected and no relevant or requested material may be withheld or destroyed. At trial, judges frown severely on plaintiffs or defendants who cannot or did not produce relevant records—and any evidence that records were tampered with or destroyed in anticipation of a lawsuit can be met by a judge with a contempt citation and a serious fine. The judge could also disallow any consideration of evidence based on the unproduced evidence. This could mean the difference between a finding for or against the party.

FIGURE 4-3 The discovery process often involves in-depth analysis of documents.

Phase 4: Settlement or Trial Preparations

Once the discovery process is complete, either side may file a motion for summary judgment. This motion states that after reviewing all the facts presented during discovery, the judge determines that no reasonable jury could find in favor of the nonmoving party. The matter is often disposed of at this stage, and just as frequently the granting of the motion is overturned on appeal. The standard for granting the motion is that there are no facts in dispute. Disputed facts are for a judge or jury to decide and the motion must be denied.

If the motion is denied, the plaintiff, defendant, and their attorneys can either settle or organize for trial. Settlement is the result of an assessment of the case and its projected costs between the defendant and its liability insurance company. If the parties settle, the lawsuit process ends here.

Phase 5: Trial

Negligence cases can be tried in front of a judge alone or in front of a judge and jury. Generally, a choice between a judge or jury trial can be made by the parties. In some cases and in some jurisdictions, the type of case and the amount of damages sought will steer the case toward a judge alone or a judge and jury. The judge acts as a combination of referee and legal scholar, overseeing the order of the trial, responding to objections made by the attorneys, and explaining the law to a jury (if there is one). When there is a jury, the members of that jury have two responsibilities: to learn the facts of the case and, after instruction from the judge, to apply the law to the facts and reach a verdict.

The jury learns the facts through a carefully choreographed presentation of electronic and paper documents and testimony of witnesses. The plaintiff presents his or her case first; the plaintiff's attorneys present documents and call witnesses. Defense attorneys, who, through the discovery process, are generally familiar with the plaintiff's witnesses and evidence, have the opportunity to object to introduction of evidence or testimony. If the evidence is from a secondary witness, the challenge might be hearsay or best evidence. In court, the best, most direct evidence is expected; other testimony or documents may be challenged as to relevance (meaningful connection between the evidence and the complaint). The judge makes the call on admission or exclusion of evidence.

When a plaintiff's witness takes the stand, the plaintiff's attorney starts the testimony process by asking direct, open-ended questions, such as, "And then what happened?" Leading questions are not allowed at this point. This phase of testimony is called direct examination. When the plaintiff's attorney has no more questions, the defendant's attorney takes a turn with cross-examination, during which the defense attorney may ask questions (including leading questions, like the classic "Isn't it true that . . . ?"), but the questions may be based only on what came up on direct examination. If the cross-examination has raised issues, the plaintiff's attorney can question the witness again in a process lawyers call *redirect*. If redirect raises more issues, the defendant's attorney can recross. Beyond this point, testimony can begin to look like a tennis match, but experienced attorneys generally do not need to go beyond the recross phase. Either side can recall a witness if subsequent testimony raises issues that the earlier witness can address.

When a case involves technical matters that are not generally known by the public (or a jury), each side can call expert witnesses, who can explain complex issues and, under some circumstances, offer an opinion about part of (or the entire) dispute.

In a negligence lawsuit, a major goal of lawyers on both sides is to make sure the jury understands the standard of care that a fire fighter or officer owes to the people he or she protects. The search for this standard is often framed in the form of a hypothetical question: What would the reasonable fire fighter (or fire officer) have done (or not done) under the same or similar circumstances? The search for evidence regarding the appropriate standard of care against which to measure the fire fighter's act or omission will begin with the fire department. Does the department have an applicable standard operating procedure (SOP) or standard operating guideline (SOG)? If yes, did the fire fighter/officer follow it? If not, lawyers will then look to the fire service community for guidance. They will ask neighboring departments (or regional or state training agencies or state fire associations) if there is an applicable standard. Finally the lawyers will look to national standards, SOPs, or SOGs (e.g., NFPA standards and National Fire Academy publications). But the lawyers do not decide if what they present is, in fact, the applicable standard. That is up to the jury.

Once the plaintiffs and defendants have introduced all their evidence, each side undertakes the task of summarizing for the jury (or for the judge sitting without a jury) the evidence they have seen and heard, analyzing and interpreting it, and explaining how that evidence supports a verdict in favor of their client. If there is a jury, it is the judge's responsibility to explain the law to the jurors (these explanations are called instructions). The jury will then consider the law and the facts (deliberate) and reach a verdict.

Appeals

Lawsuits are zero-sum events—there is a winner and a loser—and the loser can appeal. Only rarely will the losing side appeal the facts (although sometimes appeals judges will review the facts). Generally, appeals are based on the law and how it is applied (or misapplied) by the judge—evidence or testimony is admitted when it should not have been, or the judge's explanation of the law to the jury was wrong, etc. As a rule, appeals courts work hard to narrow the legal question(s) on appeal as much as possible, since an appeals court (or Supreme Court) ruling is precedent—the ruling becomes binding on trial courts in future cases. A future appeals court can overrule an earlier precedent, but because the American legal system's credibility relies in part on the consistency of the law, appeals judges infrequently do so. Depending on the legal question, a legislature can overturn an appeals court decision by statute, but because legislators share the courts' interest in consistency, a statutory overturn is likewise infrequent.

Legal Trends

As Yogi Berra is supposed to have said, "It's tough to make predictions, especially about the future." But, by its nature (and with few exceptions), the law evolves slowly enough that we can see it coming—assuming we are paying attention. Evolution in the law has created a fire service that a chief officer of 1960 would not recognize. For example, we previously discussed the changing law with regard to treatment of disabled individuals. Use of equipment that is adaptable for both disabled and nondisabled employees is becoming more prevalent.

The introduction of females into the fire station has caused an evolution in policies and procedures based on the Civil Rights Act of 1964 and subsequent amendments. Some departments may choose to ignore the wave of change, but they do so at their peril. Change also exists in the evolution of the body of law surrounding labor/management relations.

Technology

The adoption of new rules of electronic evidence has made records management and records retention more critical than ever. The law is also coming to terms with social media, use of cell phone cameras and "helmet cams," and similar technologies. On-duty use of cell phone cameras and helmet cams can be controlled by the employer. A fire fighter generally has a limited right of privacy in connection with use of the department's computer, pager, and messaging systems—that is, unless that fire fighter can demonstrate that he or she had a reasonable expectation of privacy, as the Supreme Court noted in *City of Ontario, CA v. Quon*, 130 S.Ct. 2619, 560 U.S. 746 (2010).

In general, statements or presentations by fire fighters that would be in violation of a fire department policy if uttered or presented in the workplace are equally in violation if written or presented, on duty or off duty, via a personal electronic device.

Other electronic devices are intruding in the workplace and generating liability exposure—those in the cab of a fire truck or ambulance or in the form of a cell phone used by the driver or officer on a rig. For example, in Los Angeles County, California, in 2014, a sheriff's deputy who was typing on his computer missed a curve and fatally struck a bicyclist. The department has since tightened its rules on the use of in-car computers.

Fire Officer III and IV

Laws Regarding Human Resources

Good policies and procedures must be grounded in compliance with all applicable legal requirements. These legal requirements come in the form of federal, state, and local laws and, if applicable, the collective bargaining agreement for the authority having jurisdiction or volunteer fire department bylaws. When writing any personnel policy or procedure, consideration of all legal requirements is essential. This knowledge is also useful as one is preparing for promotion.

Life Safety Initiatives

9. Thoroughly investigate all fire fighter fatalities, injuries, and near-misses.

Federal Laws

At the federal level, laws such as the Civil Rights Acts of 1866 and 1871, 1964, and 1991 prohibit all forms of discrimination in the workplace, including the area of promotions. The Equal Employment Opportunity Act of 1972 expands the scope of the 1964 Civil Rights Act, providing its legal coverage to almost all public and private employers of 15 or more people. The Americans with Disabilities Act (ADA) and the Age Discrimination in Employment Act (ADEA) may also affect the promotional policy and procedure. EEOC enforces many of the federal laws regarding employment issues and provides oversight and coordination of all federal equal opportunity regulations, practices, and policies.

Title VII of Civil Rights Act of 1964

Title VII of the Civil Rights Act of 1964 prohibits employment discrimination based on race, color, religion, sex, or national origin. Also, claims can be brought under 42 USCA § 1983 for deprivation of constitutional rights under color of state law. Title VII, however, is the principal federal statute of concern. Under Title VII, prohibited discrimination can arise from either disparate treatment or disparate impact (42 USCA § 2000e-5(g)(1); 42 USCA § 1981a(a)(1)).

Disparate treatment discrimination occurs when a member of a protected class is intentionally treated different from other employees or is evaluated by different standards. An example would be if an African American employee were disciplined for the same action for which a Caucasian employee received no discipline.

Disparate impact or adverse impact is a result when rules that are applied to all employees have a different and more inhibiting effect on a protected class than on the majority. If the organization decided that a bachelor's degree was required for a position, this could be a nonessential education requirement and have an adverse impact on a protected class. The human resources department needs to ensure that the requirements set forth are job related and part of job qualification criteria.

What this means for organizations is that any selection procedure that has a disparate impact on a protected class is invalid in predicting or measuring performance. The 4/5 rule, or 80% rule, can be used to check for disparate impact. If it is indicated that there is adverse impact, the organization has four options:

1. Abandon the procedure.
2. Modify the procedure.
3. Validate the job relatedness of the selection procedure.
4. Justify the procedure as a business necessity (Society for Human Resource Management, n.d., Module 2, 2–16).

This is why it is imperative for organizations to conduct job analyses for positions and have validated testing procedures in place.

Chief Officer Tip

4/5 Rule

1. Determine the selection rate for each group.
2. Identify the group with the highest selection rate.
3. Divide the selection rate of the group with the lowest rate by the group with the highest rate.
4. Adverse impact is present if the selection rate is less than 80%.

Much attention has been given to tests administered to job applicants. The EEOC's Uniform Guidelines on Employee Selection Procedures may require validation of testing procedure for job applicants if the test in question has an adverse impact. The EEOC urges maintenance of records that disclose by identifiable race, sex, or ethnic group the impact of applicant tests and selection procedures.

In 2012, the EEOC announced new procedures regarding the role that criminal records should (and should not) have on entry-level and promotional selection processes. Criminal-conviction statistics clearly show that a greater percentage of minorities have criminal records than whites. Therefore, screening applicants on the basis of their criminal record is likely to have a "disparate impact" on minorities. This then requires the employer to prove that screening out certain criminals is job related and consistent with business necessity.

The EEOC does not require that anyone with a criminal record be hired or promoted. The EEOC *does* require that employers consider criminal records on an individual, case-by-case basis, taking into account the following:

- The nature and gravity of the applicant's offense or offenses
- The time that has passed since the offense and the completion of any prison sentence or probation
- The nature of the job that the applicant is seeking

EEOC has advised that, regardless of state laws that contain blanket prohibitions of persons convicted of felonies holding certain jobs or certifications (fire fighter, EMT/paramedic, etc.), EEOC will maintain complaints and litigation on an employer-by-employer basis. As of this writing, there have been no definitive court decisions on EEOC's criminal-record standard.

In mid-2015, EEOC announced its conclusion that Title VII of the 1964 Civil Rights Act forbids sexual orientation discrimination on the job because it is a form of sex discrimination, which is explicitly forbidden by Title VII. EEOC argues that discrimination based on gender stereotyping, including gender stereotypes evidenced by anti-gay comments, is sex discrimination within the meaning of Title VII. As in the case of criminal records, as of this writing, there have been no definitive court decisions on this EEOC interpretation of Title VII. It is important for chief officers to remember that, in many states (perhaps their own), state laws directly prohibit discrimination on the basis of sexual orientation or gender identification.

Age Discrimination in Employment Act

The ADEA prohibits the discrimination in employment for individuals over the age of 40.

> It shall be unlawful for an employer—
>
> (1) to fail or refuse to hire or to discharge any individual or otherwise discriminate against any individual with respect to his compensation, terms, conditions, or privileges of employment, because of such individual's age; (2) to limit, segregate, or classify his employees in any way which would deprive or tend to deprive any individual of employment opportunities or otherwise adversely affect his status as an employee, because of such individual's age; or (3) to reduce the wage rate of any employee in order to comply with this Act. (29 USCA § 623)

However, the ADEA permits age discrimination "where age is a bona fide occupational qualification reasonably necessary to the normal operation of the particular business" (29 USCA § 623(f)(1) [1967]; *U.S. E.E.O.C. v. City of St. Paul*, 671 F.2d 1162, 1164 [8th Cir. 1982]).

In *U.S. E.E.O.C. v. City of St. Paul*, the court concluded that:

> [T]he district court's determination that fitness for the duties of district fire chief can be ascertained more reliably by means other than age is not clearly erroneous. Such a finding means that age cannot be said to be a "bona fide occupational qualification reasonably necessary to the normal operation" of the fire department as respecting district fire chiefs. We note, however, that our decision does not prohibit consideration of age in determining fitness for duty. It merely prohibits making age the only factor. (*U.S. E.E.O.C. v. City of St. Paul*, 671 F.2d at 1168)

Americans with Disabilities Act

Under the ADA, "[n]o covered entity shall discriminate against a qualified individual on the basis of disability in regard to job application procedures, the hiring, advancement, or discharge of employees, employee compensation, job training, and other terms, conditions, and privileges of employment" (42 U.S.C. § 12112(a)). A qualified individual is "an individual who, with or without reasonable accommodation, can perform the essential functions of the employment position that such individual holds or desires" (42 USCA § 12111(8)).

The statute itself provides that "consideration shall be given to the employer's judgment as to what functions of a job are essential . . ." (42 USCA § 12111(8)). The statute also expressly provides that "if an employer has prepared a written description before advertising or interviewing applicants for the job, this description shall be considered evidence of the essential functions of the job" (42 USCA § 12111(8)). In addition to containing a list of medical requirements that a physician may utilize in evaluating a candidate for the position of fire fighter, NFPA 1582, *Standard on Comprehensive Occupational Medical Program for Fire Departments*, also contains a list of 13 essential job tasks that a fire fighter may have to perform. In the hiring of a fire fighter, many agencies have developed a physical ability test to assist in determining whether a candidate can meet the essential job functions. One such test was developed through a joint effort between the International Association of Fire Chiefs (IAFC) and the International Association of Fire Fighters (IAFF). The Candidate Physical Ability Test (CPAT) can be a legitimate tool for assessing fire fighter employment eligibility. CPAT also meets the validity criteria established by the EEOC, the U.S. Department of Justice, and the U.S. Department of Labor, increasing its use as a legally defensible tool.

The ADA does not force an organization to make employee accommodations that would cause an undue hardship on the organization.

The basic understanding of these laws serves as a guide to asking the right questions of legal professionals and employment law experts. To gain a better understanding of each of these laws, the chief officer can review the case law and ask questions. As with all laws, different cases will challenge the current interpretation, forcing a change in how an organization needs to conduct business. The laws that govern the promotion and hiring process are only one piece of ensuring that the promotion process produces the best possible candidate.

Fair Labor Standards Act

The Fair Labor Standards Act (FLSA) establishes minimum wage, overtime pay, record keeping, and youth employment standards affecting full-time and part-time workers in the private sector and in federal, state, and local governments. Special rules apply to state and local government fire protection and law enforcement activities, volunteer services, and compensatory time off instead of cash overtime pay.

One of the primary impacts is the organization's decision regarding who is an exempt employee and who is a nonexempt employee. It is not the title that determines whether the employee is exempt or nonexempt; it is the job duties and responsibilities. The primary concern for exempt employees is that they are not entitled to overtime compensation; however, they are still covered under the other provisions of FLSA.

Another concern covered under the FLSA is when an employee performs two jobs with the same employer, for example, a full-time employee working for the Department of Public Works who is also a part-time or volunteer fire fighter. A review of the Act is required to determine if the hours worked for both services need to be combined for the purposes of computing overtime hours worked and the appropriate wage rate for the employee's overtime compensation. A misunderstanding persists in the fire service that an FLSA-exempt employee may not be suspended without pay. In 2012, the Department of Labor reiterated its opinion that DOL regulations allow for salary deductions for "unpaid disciplinary suspensions of one or more full days imposed in good faith for infractions of workplace conduct rules" "imposed pursuant to a written policy applicable to all employees." But this DOL rule is a good example of the continuing necessity to check state law as well as federal law. Some state fair labor compensation laws limit or prohibit such disciplinary suspensions.

Family and Medical Leave Act

The Family and Medical Leave Act (FMLA) is designed to allow an employee to take up to 12 weeks of unpaid leave during a 12-month period to care for a family member who is suffering from a serious health condition, for the arrival of a new child (birth or adoption), or if the employee has his or her own serious health condition. When the employee returns to work, he or she must be allowed to return to the same position or a position equal in compensation, benefits, and other conditions of employment. Federal, state, and municipal governments are required to comply with FMLA, as are private companies with 50 or more employees. Employees must have worked for the covered employer for at least one year and for at least 1250 hours in those 12 months. There are several options for the employer to define the 12-month period; however, whatever method is chosen, it must be used for all employees. The employer may opt to have the employee use paid accrued leave as part of the FMLA leave. The employee does have the option of using paid accrued leave in lieu of FMLA leave but must follow the employer's leave policies. Refer to the U.S. Department of Labor for additional information or the FMLA check sheet FIGURE 4-4. Each organization should develop its own check sheet to meet its needs.

Pregnancy Discrimination Act

The Pregnancy Discrimination Act (PDA), first passed in 1978, prohibits discrimination based on pregnancy when it comes to any aspect of employment, including hiring, firing, pay, job assignments, promotions, layoff, training, fringe benefits (such as leave and health insurance), and any other term or condition of employment.

According to the EEOC's interpretation of the Act, if a woman is temporarily unable to perform her job due to a medical condition related to pregnancy or childbirth, the employer or other covered entity must treat her in the same way as it treats any other temporarily disabled employee. For example, the employer may have to provide light duty, alternative assignments, disability leave, or unpaid leave to pregnant employees if it does so for other temporarily disabled employees.

But not all employers—or courts—agreed with the EEOC's interpretation. Only in 2015, in *Young v. UPS* (135 S.Ct. 1338 [2015]) did the U.S. Supreme Court rule directly on the question of accommodation of a pregnant employee, Peggy Young. She sued her employer, UPS, for failing to provide any accommodation for her during her pregnancy. UPS argued that the PDA and ADAAA did not require workplace pregnancy accommodation. Writing for the majority, Justice Stephen Breyer found that, "Viewing the record in the light most favorable to Young, there is a genuine dispute as to whether UPS provided more favorable treatment to at least some employees whose situation cannot reasonably be distinguished from Young's. She should be allowed to go back to court to argue that the reason she was not accommodated was her pregnancy."

Inconvenience or expense is *not* a legitimate reason for an employer to fail to accommodate pregnancy or related conditions. The Supreme Court majority also said that courts could consider (1) whether the employer made accommodations in other types of cases but not pregnancy and (2) whether the employer had multiple accommodation policies while having nothing for pregnancy.

As of January 2015, the following states currently require some form of pregnancy accommodation: Alaska, California, Connecticut, Delaware, the District of Columbia, Hawaii, Illinois, Louisiana, Maryland, Minnesota, New Jersey, Texas (government employees only), and West Virginia. As of this writing, several other states are considering enacting pregnancy-accommodation laws.

Nursing Mother Amendment to FLSA

Since the 2010 passage of the Nursing Mother Amendment to the FLSA, many employers have struggled with their legal obligation to accommodate nursing mothers—including fire fighters—who need to express milk during the workday. In 2014, Tucson fire fighter Carrie Clark brought suit against Tucson for failing to meet the requirements of the Nursing Mother Amendment. Clark returned to work as a "swing paramedic" after giving birth in 2012. (Swing paramedics work at different stations based on where they are needed.)

Not long after her return, Clark requested a transfer to a station that already had an employee pumping breast milk and contained the appropriate private area for pumping and adequate refrigerator space for storing it, court records said. Clark found a colleague who was willing to transfer out of that station so Clark might get reassigned there, but Tucson Fire Department officials ignored the request, according to Clark's suit. Beginning in January 2013, Clark began a station-to-station rotation, including stations where, she

The first revision to the Family and Medical Leave Act regulations since enactment of the 1993 law was published by the Department of Labor's (DOL's) Wage and Hour Division in the Federal Register on Monday, November 17, 2008.The following form is a checklist based on the proposed final regulations. Please note that this form is subject to change once the DOL issues final regulations. The forms referenced below are to be used in conjunction with the revised FMLA regulations effective January 16, 2009.

Note to Employers: This checklist serves as a basic tool for an employer's determination and administration of FMLA leave for an employee under the federal law and regulations. If your state's family and medical leave provisions differ from the federal law and regulations, you may need to complete additional steps for the determination and administration of family and medical leave. For complex FMLA situations and determinations . . . employers [should] obtain legal counsel review.

1. ***EFFECTIVE 01/16/09:* COVERED EMPLOYERS MUST POST THE REVISED FMLA POSTER "EMPLOYEE RIGHTS AND RESPONSIBILITIES UNDER THE FAMILY AND MEDICAL LEAVE ACT (AS REVISED 01/16/09)." DOL PROPOSED NOTICE.**
   http://www.dol.gov/whd/regs/compliance/posters/fmlaen.pdf

   ARE YOU AN EMPLOYER COVERED BY THE FMLA?
   An employer covered by FMLA is any person or entity engaged in commerce or in any industry or activity affecting commerce that employs 50 or more employees for each working day during each of 20 or more calendar workweeks in the current or preceding calendar year.
2. ***WITHIN FIVE BUSINESS DAYS AFTER AN EMPLOYEE HAS INFORMED YOU OF THE NEED FOR LEAVE, THE EMPLOYER MUST COMPLETE AND PROVIDE THE EMPLOYEE WITH THE NOTICE OF ELIGIBLITY AND RIGHTS & RESPONSIBILITIES. DOL PROPOSED NOTICE.***
3. ***ATTACH TO THE NOTICE THE APPROPRIATE CERTIFICATION FORM (ONE OF THE FOLLOWING):***
   Certification of Health Care Provider for Employee's Serious Health Condition. DOL Form WH-380-E http://www.dol.gov/whd/forms/WH-380-E.pdf
   Certification of Health Care Provider for Family Member's Serious Health Condition. DOL Form WH 380-F http://www.dol.gov/whd/forms/WH-380-F.pdf
   Certification of Qualifying Exigency for Military Family Leave. DOL Form WH-384 http://www.dol.gov/whd/forms/WH-384.pdf
   Certification for Serious Injury or Illness of Covered Service Member for Military Family Leave. DOL Form WH-385 http://www.dol.gov/whd/forms/WH-385.pdf
4. ***THE EMPLOYER MUST GIVE THE EMPLOYEE AT LEAST 15 CALENDAR DAYS TO RETURN THE FORM. ADDITIONAL TIME MAY BE ALLOWED IN SOME CIRCUMSTANCES.***

   CONSIDER THE FOLLOWING IN COMPLETING THE NOTICE TO THE EMPLOYEE:
   EMPLOYEE NOTICE TO EMPLOYER
   When the need for FMLA leave is foreseeable (i.e., childbirth or adoption), generally the employee must provide the employer with at least 30 days' notice. When an employee becomes aware of a need for FMLA leave less than 30 days in advance, it should be practicable for the employee to provide notice of the need for the leave either the same day or the next business day. When the need for FMLA leave is not foreseeable, the employee must comply with the employer's usual and customary notice and procedural requirements for requesting leave, absent unusual circumstances.
   Describe employee's notice: ______________________________
   ____
   ____
   ____

   IS THE EMPLOYEE ELIGIBLE FOR FMLA LEAVE?
   An eligible employee is an employee of a covered employer who has been employed by the employer for at least 12 months AND has worked at least 1,250 hours (actual hours worked) during the 12-month period immediately preceding start of FMLA leave, and is employed at a work site where 50 or more employees are employed by the employer within 75 miles of that work site. The 12 months of employment need not be consecutive; employment prior to a continuous break in service of seven years or more need not be counted. The employee does not have to work at a work site with 50 or more employees; the employee could work at a site with only 25 employees, for example, and still be covered by the FMLA if other employer sites within the 75-mile radius employed at least 25 employees.
   *Employee is eligible under these criteria **(circle one)**: Yes No*

   CIRCUMSTANCES REQUIRED FOR FMLA LEAVE
   Eligible employees are required to be granted FMLA leave for:
   1) The birth of a son or daughter and to care for the newborn child.
   2) For placement with the employee of a son or daughter for adoption or foster care.
   3) To care for the employee's spouse, son, daughter, or parent with a serious health condition.

FIGURE 4-4 FMLA check sheet (*continues*)

4) For the serious health condition of the employee that makes the employee unable to perform the functions of his or her job.
5) A covered family member's active duty or call to active duty in the National Guard or reserves in support of a contingency operation.
6) To care for an injured or ill covered service member.

AMOUNT, DURATION AND TYPES OF LEAVE

An eligible employee is entitled to take up to 12 weeks of FMLA leave during a 12-month period for circumstances 1 through 5 above and up to 26 weeks of FMLA leave during a 12-month period for circumstance 6 above. The employer's FMLA policy should specifically state which one of the following methods it follows for the use of 12 weeks of FMLA leave during a 12-month period.

The options are:

- The calendar year.
- Any fixed 12-month period, such as a fiscal year or a year starting with the employee's anniversary date.
- The 12-month period as measured forward from the date the employee's FMLA leave first begins.
- A "rolling" 12-month period measured backward from the date an employee uses any FMLA leave.

For the use of 26 weeks of FMLA leave to care for an injured or ill covered service member, the 12-month period begins on the first day the eligible employee takes FMLA leave to care for a covered service member and ends 12 months after that date.

Has the employee used FMLA leave in the 12-month period as described in the employer's FMLA policy ***(circle one)****:*
Yes No
If "yes" amount of leave remaining: ________________________
Expected duration of leave: ________________________
Type of leave ***(circle one)****: Continuous Intermittent Reduced Work Schedule*

MEDICAL CERTIFICATION

An employer may require that an employee's leave to care for his or her own serious health condition, for the seriously ill spouse, son or daughter, or parent, for the qualifying exigency for military family leave, or for the serious injury or illness of a covered service member for military family leave be supported by certification.
Employer requires certification under its FMLA policy or practice ***(circle one)****:*
Yes No

BENEFIT—HEALTH INSURANCE CONTINUATION AND PROTECTION

During FMLA leave, the employer must maintain the employee's coverage under any group health plan at the same level and under the same conditions as would be maintained had the employee continued actively working. The employer is required to continue its same portion of premiums as it paid during active employment.
Employee is enrolled in employer's group health plan ***(circle one)****: Yes No*
If an employee premium contribution is required, describe how employee will continue this contribution while on FMLA leave: ________________________

5. ***WITHIN FIVE BUSINESS DAYS AFTER AN EMPLOYEE HAS SUBMITTED THE APPROPRIATE CERTIFICATION FORM, THE EMPLOYER MUST COMPLETE AND PROVIDE THE EMPLOYEE WITH THE DESIGNATION NOTICE. DOL FORM WH-382 EMPLOYER RESPONSIBILITY TO DESIGNATE FMLA LEAVE AND NOTICE TO EMPLOYEE***
http://www.dol.gov/whd/forms/WH-382.pdf
In all circumstances, it is the employer's responsibility to designate leave, paid or unpaid, as FMLA-qualifying and to give notice to the employee. If the employee refuses to provide required information and the medical certification under the employer's FMLA policy, the employer would not be able to designate the absence as FMLA leave and the employee would not have the FMLA protections and continuation of benefits of FMLA.

FIGURE 4-4 FMLA check sheet (*continued*).

alleged, there were inadequate or no facilities available for her to pump breast milk. As of this writing, Clark's suit has not proceeded through litigation.

State Laws

At the state level, too, there are laws concerning equal employment opportunity issues. Some states may even have specific promotion legislation. For example, the state of Illinois recently passed into law the Fire Department Promotion Act (Illinois Compiled Statutes, 50 ILCS 742). The Act covers such topics as the following:

- Applicability (positions to which the act applies)
- Promotion process
- Promotion lists
- Monitoring
- Written examinations
- Seniority points
- Ascertained merit

- Subjective evaluation
- Veteran's preference
- Right to review

As with the federal government, the states also have enforcement and regulatory agencies. These state agencies are called human resource commissions, commissions on human relations, fair employment commissions, state department of personnel, state department of labor, and so on.

Records Retention and Public Access

While federal laws governing the retention of federal records and public access to those records have been passed, chief officers must review these issues on a state-by-state basis. Each state has laws governing the retention of public records, as well as their own version of the federal Freedom of Information Act (FOIA).

Throughout the normal course of business, fire agencies generate many different types of records. From incident reports to apparatus inspection records, these documents must be maintained according to applicable state laws. Records retention schedules established through state law are developed to guide public officials in the storage and purging of all public documents, including fire service records. For example, some incident reports might be required to be maintained for a set period of years before they are allowed to be destroyed. Missouri state law says that fire investigations, fire incident/non-fire emergency reports, and fire incident indexes must be retained for a period of 5 years for minor fire and non-fire emergencies and 50 years for major fires and losses due to non-fire emergencies. Records should also be evaluated for historical purposes. In Michigan, records that document a response to an incident by fire/EMS/ambulance must be kept for the current period plus an additional 10 years.

Chief officers should also investigate state laws regarding how records are kept. Some best practices indicate that if records are retained in electronic format they should be periodically migrated and transferred to a more permanent record format such as microfilm.

In addition to retaining records, state laws also govern public access to those records. Through FOIA, the public has a right to access for viewing and/or copying certain public records. Specifics of FOIA laws can vary from state to state. The National Freedom of Information Coalition website contains information on state FOIA laws. Because time limits for responding to public FOIA requests, amounts that a government agency can reasonably charge to process a request, and which documents are exempt from FOIA requests vary from state to state, chief officers need to review their state's statutes to be sure they process FOIA requests according to their state requirements. Additional information on records retention can be found in the "Communications" chapter.

Local Laws

The local governing body of the municipal fire department, fire protection district, or volunteer department may also have legal requirements for promotion. These requirements are generally well known within a department, but they should be reviewed for compliance with federal and state laws before incorporating them into a promotional policy or procedure.

Employee Accommodation

It is not possible to overstate the importance of the "reasonable accommodation" process for ADA/ADAAA (and when religious practices, for example, clash with workplace requirements). A reasonable accommodation is the result of an interaction between the employer and employee where:

- The job's essential functions are revisited
- Consultation with the employee's medical representative can take place
- Potential accommodations are identified and examined
- An accommodation is identified that best suits the needs of the employee and employer.

It is important to note that a mutually acceptable employee accommodation may not result from this process—the accommodation desired by the employee may result in undue hardship on the employer, or the employer's best accommodation option may not be acceptable to the employee. In either case, should the employee decide to challenge the employer's decision, EEOC (or the court) will be looking for evidence that the employer and employee made a reasonable accommodation effort and will examine the reasonableness of the employer's and employee's accommodations offers. In the absence of this accommodation process, the employer will almost always lose an ADAAA (or First Amendment) challenge.

Examples of accommodation include:

- Unpaid leave
- Modified work schedule (for example, shift-swapping, when practical, for a day away from work for a religious holiday)
- Light duty

Note that, under federal law, light duty is an accommodation and not a requirement, and an employer can—to a limited extent—maintain a policy that reserves light duty for employees only when they're injured on duty. But the EEOC's position regarding light duty for pregnant employees under the PDA and ADAAA is that if a light-duty accommodation is made for anyone, for any reason, it must also be available as an accommodation option for the pregnant employee.

The accommodation process can be complex; best practice is consultation with an attorney familiar with the accommodation process.

After the ADAAA in 2008, many of the court decisions that acted as limitations on the provisions of the original act were repudiated by Congress. The Act's remedies are now available to a much broader class of individuals who were previously excluded by the court cases.

Additional information on the ADA can be found online. The ADA was amended in 2008 and is now known in the legal community as ADAAA (ADA as amended). The principal amendment was a relaxation of the "permanent impairment" standard set by the U.S. Supreme Court in *Toyota*

Motor Manufacturing, Kentucky, Inc. v. Williams, 534 U.S. 184 (2002), and a general direction by Congress that ADAAA was to be broadly construed in favor of the disabled employee. In *Summers v. Altarum Institute Corp.*, 740 F.3d 325 (2014), the Fourth Circuit Court of Appeals ruled that a severe but temporary back injury was a disability for purposes of ADAAA, and the disabled employee was thus protected by ADAAA. In the words of the EEOC guidance on ADAAA, "although impairments that last only for a short period of time are typically not covered," they may be covered "if sufficiently severe."

Chief Officer Tip

Job Descriptions for All

A job description for every position in the department is essential. The job description should not be too short; it must include detailed descriptions of all requirements of the position that are essential to the performance of the job. Details include distances traveled and means of travel, specific equipment used, height of lifts or climbs, weight carried or manipulated, weather conditions, etc. The job description must apply to all of the persons in the position and their ability to perform those functions.

Justifying an Accommodation

The first requirement in justifying an accommodation is a request by the individual for accommodation. The second requirement is a physician's certification of the condition, including whether the individual can perform the essential functions of the position with accommodation. If possible, the physician should provide suggested means of accommodation, but doing so is not required.

Once the ability to perform the essential functions is determined and a need to provide an accommodation is identified, it is important to provide adequate information to justify the accommodation as reasonable. To make a well-informed decision, justification of the request is required by the person or persons in the department who have the authority and responsibility to approve the accommodation.

The justification should include the requirements of applicable law and the rationale for an accommodation outside the parameters of the law. Compliance with any bargaining unit contract language and department policies or rules and regulations should be considered as well. Supportable financial consequences of the accommodation should also be included.

Accommodation is required unless the employer can show that accommodating the individual would pose an undue hardship on the employer. Undue hardship refers to any accommodation that would be unduly costly, extensive, substantial, or disruptive or that would fundamentally alter the nature or operation of the business (EEOC 2011). Examples would be changing the work shift to accommodate someone who could not see in dim light or providing specially purchased equipment that has an exorbitant cost. An example of what would not be considered an undue hardship is providing a face piece that can accommodate prescription vision corrective lenses. Additionally, in the case of a nonemergency job, such as fire inspector, the employer may consider providing time off for medical treatment, adjusted work schedules, nonconsecutive shift hours, or even specific occupancy inspections, if available. The determination is extensively fact based and must be reviewed with an attorney before a decision is made.

Plan Implementation

A plan for implementing the accommodation is a necessary step in fulfilling the accommodation's intent. First, identify the specific needs of the person being accommodated. Then, determine how best to accommodate the specific needs. This may entail specialized equipment such as the face piece that can accept prescription lenses; a modified workspace, such as higher or lower desk tops or chairs; special training, such as individual equipment familiarization; and a modified work schedule, such as time off for medical treatment. Other details such as budget requirements and a timeline for implementation should be included in the plan as well. The governing board of the unit of local government will be the final approval authority. Once approved and appropriated, the plan should be implemented without delay.

Providing an employee accommodation can be a difficult undertaking due to the culture of the fire service in general and the culture of public finance in particular. There may be significant financial costs involved and the accommodation may be viewed by some as unnecessary or inconsistent with past practice, such as elevators to allow public attendance at meetings (or moving the meeting to another location to permit public attendance). However, providing the accommodation allows for compliance with a legal obligation and allows a member to remain or become a valuable asset of the department. It should be noted that there are accommodation requests that are unreasonable and need not be provided; however, such circumstances are fact specific, and consultation with a firm that is well versed in the requirements of the act is essential.

State Laws for Personnel Relations

State personnel laws often make provision for public employee rights and benefits, including statutes that govern disciplinary action and termination. (Union contracts also include provisions on discipline and other personnel matters.) Some states have enacted laws that regulate hiring new fire fighters. In other states, the local employing government regulates public employee issues via local ordinance.

Most states have adopted their own parallel versions of federal laws protecting individual rights. These parallel laws include state versions of the Civil Rights Acts of 1964 and 1991, ADA, and FMLA.

A recent trend in state and local employment law involves the use of alternative dispute resolution (ADR) methods as adjuncts to (or substitutes for) traditional disciplinary and grievance systems. Mediation involves a neutral third-party mediator who meets with both parties of the dispute and, through careful listening and structured questioning, the parties are encouraged to develop their own resolution of the complaint. The U.S. Postal Service slashed its grievance rate when it adopted mediation as an alternative to the grievance process (at the employee's option).

A second example of ADR is arbitration. An arbitrator, unlike a mediator, hears evidence and renders a decision. Arbitration is most often associated with resolving complex disputes, like collective bargaining disputes. In most states, the law involving public safety employee contract disputes requires binding arbitration in exchange for the statute prohibiting public employee strikes.

Design Credits: Flames: © Photos.com; Chapter opener: © Jason Reed/Reuters/Landov; You Are the Chief Officer: © Jones & Bartlett Learning. Photographed by Kimberly Potvin; Voices of Experience: © Glen E. Ellman; Smoke: © Greg Henry/Shutterstock, Inc.; Chief Officer in Action: Courtesy of Scott Dornan, ConocoPhillips Alaska.

You Are the Chief Officer Summary

1. What are the essential functions of the position of fire prevention inspector without emergency response duties?

Among those items to be considered are:

- Blueprint reading
- Distance measuring
- Ability to read a tape measure/architectural or engineer scale
- Ability to crawl in small spaces
- Ability to see in dim light
- Ability to traverse uneven, rutted, muddy, or ice-covered ground
- Ability to climb stairs to levels above and below grade
- Ability to wear safety shoes, eye protection, coat, hard hat, sound protection, and self-contained breathing apparatus

2. Your hiring process for the position of fire fighter will include a physical ability test and medical evaluation to determine fitness for duty. What guidance can you follow when developing a physical ability test that realistically assesses the physical requirements of a fire fighter, and what information is available to a physician who is evaluating a potential fire fighter who has an offer of conditional employment?

NFPA 1582, *Standard on Comprehensive Occupational Medical Program for Fire Departments*, lists the medical requirements for the position of fire fighter. These requirements can be used by a physician when evaluating a fire fighter candidate. NFPA 1582 also lists 13 essential job functions that are typical of tasks a fire fighter may be asked to perform. The IAFC and the IAFF have jointly developed a Candidate Physical Ability Test (CPAT) for use in assessing whether candidates are physically capable of performing essential job tasks.

3. On the application for employment as a fire fighter, what questions should not be included? Why?

Questions regarding age (except to determine whether the applicant is over the statutory age limitations that apply in the jurisdiction), racial/national origin (discrimination), marital status (irrelevant), workers' compensation history, arrest record (the question regarding convictions is for the background check), credit/bankruptcy (for the background check), and any medical/genetic information (including height and weight—permitted only after a bona fide job offer).

4. You receive a request for accommodation by a dispatch applicant who is farsighted and requires significant magnification and light for reading. How do you determine whether the request is unduly burdensome?

Determine what lighting changes would need to be made in the work area and what they would cost. Can a magnifying lamp be provided? Can an enlarged computer monitor be provided? Can enlarged-print documents be provided? Is the cost of these changes too much?

Wrap-Up

Chief Concepts

- Chief fire officers must be ever aware of the sometimes complicated legal issues that affect our decisions and actions on a daily basis.
- The American fire service is governed by state and local statutory and common law.
- North American settlers brought their law with them. To improve the federal government, the states departed from the British model of common-law rights and responsibilities of citizens and the monarchy and adopted a written constitution in 1787.
- Good policies and procedures must be grounded in compliance with all applicable legal requirements. When writing any personnel policy or procedure, consideration of all legal requirements is essential.
- To recover damages from negligence, each of the following elements must be proven:
  - Duty to act
  - Breach of duty to act
  - Proximate cause
  - Actual injury
- The objective of the American legal system is to handle disputes like negligence claims by mediation, arbitration, or other settlement methods short of a lawsuit. If a dispute cannot be settled and a lawsuit is filed, the American legal system continues to encourage efforts at settlement prior to trial.
- By its nature (and with few exceptions), the law evolves slowly enough that we can see it coming—assuming we are paying attention.
- The law and the fire service often interact in connection with individual rights protected by the U.S. Constitution, state constitutions, and federal and state statutes.
- State personnel laws often make provision for public employee rights and benefits, including statutes that govern disciplinary action and termination. (Union contracts also include provisions on discipline and other personnel matters.)
- Federal laws and statutes regulate actions such as drug tests, workplace searches, and free speech.
- When just cause is required for severe discipline or dismissal of a local government employee, the court held (*Cleveland Board of Education v. Loudermill*) that the employee had a property interest in his or her job and was entitled to due process.

Hot Terms

Administrative warrant A warrant that requires only a showing of reasonableness on the part of a fire inspector to a judge.

Answer Response to the complaint.

Appellate courts Courts with precedential powers; also called courts of record; the intermediate court between the district court and the Supreme Court.

Burden of proof A procedural term that dictates which party has the obligation to prove a wrong has been committed by the defendant.

Common law Law established in appeals-level courts and relied on in future cases.

Complaint A recitation of the facts by one or more plaintiffs that creates liability on the part of one or more defendants and a request for damages.

Courts of record Courts with precedential powers; appellate courts.

Cross-examination The defendant's attorney's chance to ask questions (including leading questions) of the plaintiff's witness, but based only on what came up on direct examination.

Daubert standard Ensures that scientific evidence is relevant and reliable and that what is represented as scientific knowledge is, in fact, based on analysis that meets standards of the scientific method.

Defendant The person allegedly causing the injury.

Deliberate To consider the law and reach a conclusion regarding the law and the evidence produced during a trial.

Depose Examine under oath, regarding the witnesses' knowledge and involvement with the facts in the complaint.

Dillon's Rule Legal principle that states that a unit of local government has only as much authority as the state legislature grants to the local government or that can be implied as a result of the grant of power.

Direct examination Direct, open-ended questions the plaintiff's attorney asks the plaintiff's witness. Leading questions are not allowed at this point.

Discovery The phase in the lawsuit process in which attorneys representing the plaintiff and defendant conduct research for information relevant to the case.

Discretionary acts Local government acts requiring the exercise of judgment.

Due process The opportunity to know the charges, to be heard, and to have evidence fairly considered prior to suspension, demotion, or termination.

Employee accommodation The steps an employer can take to help a person with disabilities perform a job, provided the individual is capable of performing the essential functions of the job.

Garrity warning Verbal warning given to the subject of a disciplinary interrogation that informs the subject that he is required to answer all questions and that he is being given immunity from criminal prosecution for those answers.

Governmental activity Activity unique to government.

Ministerial acts Acts requiring little or no judgment.

Negligence Condition in which someone had a duty to act, breached that duty, that breach is the proximate cause of an injury, and that breach inflicted actual injury.

Nongovernmental fire departments Departments that operate as private, nonprofit, or for-profit corporations and contract with towns, cities, or counties to deliver services within specific boundaries.

Ordinary negligence An unintentional act that causes injury.

Plaintiff The complaining party.

Precedent The court's decision that is controlling law within that court's jurisdiction in any future case involving the same (or similar) facts.

Proprietary activities Activities undertaken by both government and the private sector.

Relevance Meaningful connection between the evidence and the complaint.

Statutory law Law adopted by legislative bodies.

Written interrogatories Questionnaires answered by potential witnesses under oath.

Walk the Talk

1. You are assembling a panel of fire service peers to assist you in interviewing candidates for a fire fighter position. Each of the four panel members will be asked to develop three questions that will be asked during the interview process. What guidance can you provide the panel members to help them develop questions that will not violate federal and/or state laws governing civil rights?
2. As a new chief officer you realize that the department has no policy for handling the department's many records. After discussions with your supervisor on this issue, you are to research your state laws governing records retention and fire department compliance with Freedom of Information Act requests. A policy is to be developed dealing with the department's handling of its records and submitted to your supervisor for review.
3. A legal trend is to hold chief officers personally liable for fire department bad outcome events, such as a line-of-duty death or a large loss fire. Should chief officers have a personal attorney to advise them on departmental issues?
4. In handling a high-profile personnel matter, a chief officer's supervisor is advocating a resolution that is not consistent with the IAFC's Code of Ethics and appears to circumvent local administrative law. What options does the chief officer have?
5. Discuss how a chief officer can stay informed on changing legal issues.

CHIEF OFFICER *in action*

A meeting of regional chief officers includes discussion of an approach by the Chicago Police Department in handling police misconduct complaints. Faced with a continuing increase of police misconduct cases overloading the court docket, the department of law was settling many cases out of court. Although this cleared up the court case workload, it severely burdened the city budget and impacted police officer morale and effectiveness.

In July 2009, the superintendent of police changed the strategy. Instead of settling cases out of court, every misconduct case would be brought to trial. The superintendent stated, "If plaintiffs know their complaint will in fact be litigated, more focus and concern will be given to the factual validity of the complaints signed." A perception was that the city was quick to settle, resulting in many frivolous misconduct complaints.

Implementation of this policy required contracting with 14 different law firms to handle the workload. Impact on filings was immediate. In Fiscal Year 2010, the number of civil rights cases filed against police officers dropped by nearly 50%, and non–civil rights cases brought against the officers were voluntarily dismissed at a higher rate. In 2009, 18% of the plaintiffs voluntarily dropped their case. By October 2010, more than 45% dropped their cases. The department of law told the city that the results are "nothing short of astounding" (Kerrigan 2011).

1. Voluntarily dismissing a complaint most likely occurs after the __________ phase.
   - **A.** complaint
   - **B.** answer
   - **C.** discovery
   - **D.** trial
2. Appeals of legal decisions are usually based on:
   - **A.** additional information.
   - **B.** economic considerations.
   - **C.** justice consequences of the original decision.
   - **D.** how the law was applied.
3. Fire Captain Doe witnessed a collision between a police cruiser and a private vehicle. His description of the event would be part of a(n):
   - **A.** interrogatory.
   - **B.** Dillon's Rule exception.
   - **C.** civil penalty determination.
   - **D.** USERRA Act activity.
4. A chief officer could be directly involved in the __________ portion of a lawsuit.
   - **A.** direct examination
   - **B.** answer
   - **C.** discovery
   - **D.** All of the above
5. Engine 1 is responding to a life-threatening emergency with all emergency lights activated. With siren screaming and air horn blowing, the rig does not slow down for a red-light-controlled intersection. To prove negligence, the injured party must show:
   - **A.** breach of a duty of care resulted in actual injury.
   - **B.** breach of a duty of care could have resulted in an actual injury.
   - **C.** the driver knowingly violated departmental regulation to come to a complete stop.
   - **D.** the driver violated state regulations.

CHAPTER 5

Human Resources

Fire Officer III

Knowledge Objectives

After studying this chapter, you should be able to:

- Explain documentation and record keeping of human resource materials NFPA 6.1.2. (pp 105–107)
- Discuss wellness and fitness initiatives for fire service employees NFPA 6.2 NFPA 6.2.2 NFPA 6.7 NFPA 6.7.1. (pp 107–110)

Skills Objectives

After studying this chapter, you should be able to:

- Compile and maintain human resource documentation NFPA 6.1.2. (pp 105–107)
- Implement wellness and fitness initiatives for fire service employees NFPA 6.2 NFPA 6.7 NFPA 6.7.1. (pp 107–110)

Fire Officer III and IV

Knowledge Objectives

After studying this chapter, you should be able to:

- Explain how to anticipate the human resource needs of a fire organization NFPA 6.2.2 NFPA 7.2 NFPA 7.2.3. (pp 110–112)
- Describe the demographic characteristics of your department and its community NFPA 6.1.2 NFPA 7.2 NFPA 7.2.1. (p 112)
- Explain how to maximize efficiency in human resources in your organization NFPA 6.2 NFPA 6.2.1 NFPA 6.2.2 NFPA 7.2.3 NFPA 7.7.1. (pp 112–117)
- Describe various employee benefits and their importance NFPA 6.2 NFPA 6.2.5 NFPA 7.2.5. (pp 118–121)
- Discuss personnel management in a fire organization NFPA 6.2 NFPA 6.2.1 NFPA 6.2.2 NFPA 7.2.2 NFPA 7.2.3 NFPA 7.2.4. (pp 121, 124–129)
- Discuss the elements of personnel promotion, including professional development and succession planning NFPA 6.2 NFPA 6.2.1 NFPA 6.2.3 NFPA 6.2.4 NFPA 6.2.7 NFPA 7.2 NFPA 7.2.3. (pp 129–136)
- Explain labor relations in the context of human resources NFPA 6.1.1 NFPA 7.4 NFPA 7.7.1. (pp 136–138)

Skills Objectives

After studying this chapter, you should be able to:

- Anticipate human resource needs in a fire organization NFPA 6.2.2 NFPA 7.2 NFPA 7.2.3. (pp 110–112)
- Perform a demographic survey of your department and its community NFPA 6.1.2 NFPA 7.2 NFPA 7.2.1. (p 112)
- Maximize efficiency in human resources in your organization NFPA 6.2 NFPA 6.2.1 NFPA 6.2.2 NFPA 7.2.3 NFPA 7.7.1. (pp 112–117)
- Implement various employee benefits NFPA 6.2 NFPA 6.2.5 NFPA 7.2.5. (pp 118–121)
- Promote personnel fairly NFPA 6.1 NFPA 6.2 NFPA 6.2.1 NFPA 6.2.3 NFPA 6.2.4 NFPA 6.2.7 NFPA 7.2 NFPA 7.2.3. (pp 129–136)

Fire Officer IV

Knowledge Objectives

After studying this chapter, you should be able to:

- Discuss the role department culture plays in the fire service NFPA 7.2. (pp 138–139)

Skills Objectives

There are no Fire Officer IV-only skills objectives for this chapter.

As you begin your third year as the fire chief, you begin to reflect on your advancement into this position and what it took for you to reach this goal. You remember the decision to go back to school and earn a degree, as well as all the time put into department activities, training, and responses. Knowing that your plan is to work 10 more years until retirement, you decide that this is an excellent time to develop and implement a department-wide succession plan. Your goal is to prepare the fire fighters, staff officers, and your administrative chiefs to be able to promote into future positions as they become open after you leave the service.

1. How would you explain the concept of a succession plan to your employees?
2. How can you and your officers recognize candidates who have the potential to fill future positions?
3. What actions can be taken to prepare individuals for future promotions?
4. In which activities would you involve your deputy chief as part of the plan to prepare him or her for the position of fire chief?

Introduction

Human resource management involves all the activities that relate to managing a department's most important asset: its personnel. Everything a department does should revolve around providing the best possible services to the citizenry it protects. Ensuring the internal customer's needs are being met allows employees to concentrate on meeting the needs of the citizens they serve. Providing the community with efficient, effective, and safe fire service quite simply cannot be accomplished without dedicated, well-trained, and highly motivated personnel.

The traditional roles of human resource managers have changed and will continue to change to meet the needs of the workforce. Human resource management has become more than overseeing records management or establishing employee compensation and benefits. This chapter discusses important administrative topics relating to human resources in any fire service organization. All departments, whether staffed with career, part-time, volunteer, or a combination of personnel types, can benefit from the information presented. The material presented in this chapter can help in the decision-making process in a variety of human resource areas, including staffing, wage and benefit packages, hiring and recruiting, management/employee relations, promotions, succession planning, training program management, and continuous improvement. As decision makers, chief officers and others in positions of authority and responsibility—such as city managers, local officials, trustees, and board members—should have a good understanding of the topics discussed in this chapter.

Fire Officer III

Record Keeping

Record keeping is an important responsibility in any organization. Records in the fire service document the major areas of department activity and function. Many record-keeping categories are mandated by law, good organizational practice, and ethics. Major categories of records include the following: financial, personnel, administrative, and emergency and nonemergency activities. Records take two primary forms—hard-copy documents and electronic records.

Proper Documentation

The proper documentation of all department activities and functions fulfills three major objectives: compliance with legal requirements, data proving well-informed decision making, and justification for resource allocation.

When developing, maintaining, or evaluating a record system, first consider what the system is supposed to provide. This is an evaluation process that answers four basic questions:

1. Is the system providing the necessary information for compliance with legal requirements and best practices?
2. Is the system providing high-quality data that support well-informed decision making?
3. What information do I want in the records?
4. In what form do I want the information?

The answers to these basic questions may identify areas that need improvement or indicate whether additional record-keeping programs need to be developed.

Completeness and accuracy are the cornerstones of any record-keeping system. Without complete and accurate records, information critical to department organization and operations is not available. Ensuring completeness and accuracy can be achieved only if good policies and procedures are in place and properly followed. Good policy ensures that the goals of the system are accurate and completely described. Good procedures ensure that the record system fulfills the intended goals by providing record keepers with good procedural guidelines.

Documentation Procedures

Procedures are not effective without good parameters for data entry or the means to enter, correlate, and distribute the required information. Proper procedures, allowing for complete and accurate data entry, should be in place. Effective means of data entry, correlation, and distribution of the information can be achieved only with a well-planned system. Good computer systems that include comprehensive and user-friendly hardware and software programs are widely available and can be used in any size department.

Once procedures and the means to comply with the procedures are in place, initial and ongoing training becomes an integral part of the system. Adequate training ensures that all personnel who enter data into the record-keeping system are proficient in providing quality data.

Records Retention

How long a record is kept may be governed by state laws. In Michigan, for example, the State Library approves and establishes a records retention schedule for all documents held by government agencies. The retention schedule for fire departments outlines which documents are kept and for how long.

Complying with a records retention schedule becomes important when dealing with Freedom of Information Act (FOIA) requests. FOIA rules were passed to improve public access to government documents and information. These laws were intended to increase the transparency of government and to provide citizens with the mechanism to request information from all levels of government: local, state, and federal. FOIA regulations spell out the time frames for responding to an information request and provide for appropriate collection of fees to cover the cost of preparing the documents. Under FOIA, citizens may request to receive copies of documents or to simply view those documents onsite. Nothing in FOIA, however, requires a government entity to create a new document. In other words, the FOIA request must identify the specific documents that currently exist and cannot request information that must be researched, collected, and produced in written form. The "Legal Issues" chapter contains additional information on this topic.

When a record is not listed in the retention schedule, departments are free to determine the length of time that document will be retained. There is nothing preventing a department from keeping records beyond the retention schedule guidelines; however, if kept, the records remain subject to FOIA requests. Chief officers should review individual state laws concerning how records are to be retained FIGURE 5-1. Questions regarding the format of the records (e.g., in original form or as microfilmed copies) should be answered at the local level. Municipal clerks are good sources of information concerning records retention and may assist in providing access to retention schedules.

FIGURE 5-1 How records are kept and how long records are retained is determined by state law.

Historical Records

Beyond the need to comply with retention requirements, some documents are retained for a historical purpose. These documents chart the history of the department and provide a timeline of important events. As an example, newspaper articles are often retained for this purpose. In today's media-driven society, recorded media such as that found on the Internet or in television newscasts may also be retained for historical purposes. Documents held for historical information may also be subject to FOIA requirements.

Analysis of Records and Data

Records and data are of little use if the record and data system cannot provide factual information. Consider the old adage "Garbage in, garbage out," in which garbage is defined as useless, inaccurate, or unhelpful data. Garbage in any data system should be identified and eliminated. A good system should provide complete and valid information that is easily retrieved and interpreted. Well-informed decision making demands good and accurate information. Many important decisions are based solely or in part on the records and data produced by the system. These decisions cover many areas of administration such as resource allocation, budgeting, and short- and long-range planning.

Validity

The first step in providing information from records and data systems is ensuring that the information provided is valid. Information that is not valid is of little or no use, even if it is the only available data. One source of data collected by departments is the reports that document department responses. Because it is not uncommon for reports to be completed by multiple individuals, issues of consistency in the reported data often arise. To ensure validity of these data, reporting policies need to ensure consistent collection, and all reports should be quality checked for compliance with reporting requirements. If the information is considered to be partially invalid, this should be noted in any analysis activities.

Interpretation

When analyzing and interpreting records and data, one must ensure that the records and data are as complete as possible given the system from which they were obtained. If the system provides partial or incomplete records and data, look for other sources that can provide similar or anecdotal information. One of the keys to well-informed decision making is gathering sufficient information, from multiple sources if available, on which decisions can be based.

Once one has gathered records and data as completely as possible, the next step is to put the information provided into a usable form that is easily understood by those who are going to use the information. Bar graphs, pie charts, line graphs, and spreadsheets are commonly used to put information in an easily understood format.

When analyzing and interpreting information, it is helpful to have an inquisitive mind-set and to evaluate for what the information is really saying. During the process of analyzing and interpreting, keep in mind that there are pitfalls in the process. It is important to have a good understanding of what is being interpreted. For example, if information on the number of structure fires is needed, statistics for annual fire calls may be misleading if information on the number of fire call responses in a particular jurisdiction is not interpreted correctly. One system may provide the number of calls classified as a fire response, but these data may not be useful or may need more separation because they may not provide data on actual structure fires that have taken place. Often fire departments respond to a fire call or structure fire call to find a minor situation that does not involve what is typically described as a structure fire.

Records and data on statistics can also be misleading if those interpreting the information do not question the data. For example, injury statistics gathered from records and data indicate a growth in the number of injury reports over the last five years at a rate of 20 percent. If call volume is consistent over the time period, it may be reasonable to assume there is an emerging safety problem; however, if the call volume has increased 25 percent over the five-year period, it may be reasonable to assume the safety record has improved slightly. Keeping an inquisitive mind-set and looking at the overall situation when interpreting records, data, and the information they provide can help to ensure that the information is useful in the decision-making process.

During the process of analyzing and interpreting records and data, note flaws in the system. Flaws might include insufficient data or information, lack of the necessary hardware or software systems that produce the records or data, inaccurate policy or procedures that cause improper or inaccurate entry of information into a system, insufficient training of those entering information into the system, lack of supporting information, poor integration with other systems, and inability of the system to produce information in its most usable form. Once any flaws are identified, the chief officer can prioritize them and undertake methods of improvement.

Employee Wellness and Fitness

It is said that statistics do not lie. If the reported numbers of fire fighter injuries and line of duty deaths are true, then chief officers do not have to look very far for statistics to justify the establishment of a wellness and fitness program. Issues regarding fire fighter wellness and fitness continue to be the number one cause of fire fighter line-of-duty deaths, with heart attacks leading the list. In addition, the wellness and fitness of employees affects multiple aspects of the organization. Increased sick leave usage, higher injury rates with corresponding increases in workers' compensation claims, possible early medical disability retirements, and increases in medical insurance claims and premiums can all result from an unhealthy workforce. Chief officers should therefore make the wellness and fitness of their employees a priority from a planning, human resource, and budget standpoint.

Wellness and fitness issues have received much attention in recent years. Departments have paid a greater attention to the health of their employees through the initiation of joint employee and management discussions and efforts. A complete wellness and fitness program can include many components and should address employee wellness from the time of recruitment until employee retirement. These components can include essential job tasks in the development of fire service job descriptions, use of a candidate physical ability screening tool, comprehensive medical examinations, increased employee involvement in health and wellness issues, convenient opportunities for fire fighters to physically exercise, and rehabilitation support for injured employees or others unable to meet the physical requirements of the job.

In 1997, two major players on the national fire service scene—the International Association of Fire Chiefs (IAFC) and the International Association of Fire Fighters (IAFF)—joined together with representatives of 10 unions and their departments to form the IAFC/IAFF Fire Service Joint Labor Management Wellness-Fitness Task Force. This partnership resulted in the publication of a manual that provides valuable information to chief officers and employee representatives seeking to establish their own wellness-fitness program.

Job Descriptions

Developing an appropriate job description for all positions within the fire department has always been an important human resource recommendation. What is included in those job descriptions is also important from a wellness and fitness point of view. A fit workforce begins with hiring fit employees.

Federal labor laws require that applicants be treated fairly and equally. That does not mean, however, that applicants cannot be screened for the ability to perform the job. This is exactly where a well-developed job description is important. Valid job descriptions should spell out what are called the essential job tasks for the position. Because the majority of those entering the fire service do so at the fire fighter level, it is important that the fire fighter's job description accurately identifies the physical requirements necessary to do the job. Hiring capable personnel at that level improves, although does not ensure, the chances that employees will remain capable of successfully and safely performing the expected fire fighter tasks. NFPA 1582, *Standard on Comprehensive Occupational Medical Program for Fire Departments*, may provide guidance when developing essential job tasks for a department's job description. While chief officers may review the 13 essential job tasks of a fire fighter included in the standard, it is recommended that a department develop its own list of essential job tasks that are reflective of their operation. For example, one department may require fire fighters to be able to climb multiple sets of stairs during a high-rise incident, whereas that requirement might not apply to a department in a rural setting. The bottom line is that for a job description to be considered valid and survive a discrimination challenge it must accurately reflect the physical nature of the actual tasks performed by the employees for which it was written.

Candidate Ability Testing

An important part of any fire fighter hiring process is the evaluation of that candidate's ability to perform the job. Fighting fire is hard work. It is stressful and requires a level of physical strength and endurance not found in many other occupations. With the position's essential job tasks included in the job description, a physical ability test component can be added to the hiring process.

Physical ability testing should be designed to test the candidate's ability to meet the physical demands of the job FIGURE 5-2. As long as the candidate can successfully complete the testing, he or she should remain in contention for the position. Ability tests become a problem for the organization when they are developed in a manner that artificially inflates the physical demands of the job. Chief officers who utilize a test that does not accurately reflect job tasks as a means to consider only those candidates with the highest level of physical strength may find themselves involved in a discrimination complaint. The IAFC/IAFF Joint Labor Management Wellness-Fitness Initiative includes a Candidate Physical Ability Test (CPAT) that can be utilized for the evaluation of fire fighter candidates. For information on CPAT and licensing requirements contact either the IAFC or IAFF.

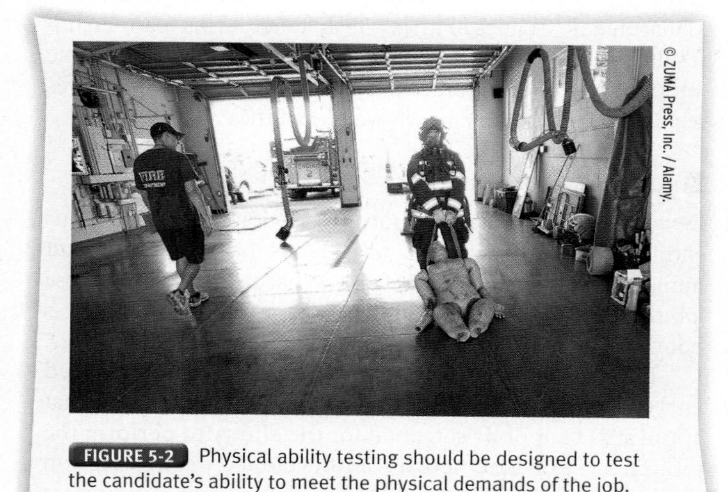

© ZUMA Press, Inc. / Alamy.

FIGURE 5-2 Physical ability testing should be designed to test the candidate's ability to meet the physical demands of the job.

Comprehensive Medical Examinations

Firefighting is not the only occupation that has a medical examination as a job requirement. Perhaps a better question is what type of medical examination should be given to potential employees. Does a medical exam designed for a foundry worker suffice for a fire fighter? Both employee groups may be required to work in extreme heat.

Implementing employee and candidate medical exams requires a partnership with the organization's healthcare provider. Although the physician is well educated in the medical field, he or she will need additional information from the organization in order to evaluate an individual for fitness for duty as a fire fighter. Medical exams are given under three pretexts:

1. First, a candidate who has received a conditional offer of employment may be asked to undergo a thorough medical exam prior to beginning work.
2. Second, an existing employee may be required to undergo an annual medical exam to ensure that the employee continues to remain fit for duty.
3. Third, an employee may be required to pass an examination before returning to duty after an injury leave.

In any case, it is the responsibility of the department to educate the medical professional on the physical and medical requirements of the position.

NFPA 1582 contains medical standards that the physician can utilize in making a proper recommendation on the medical fitness for duty of a fire fighter. While the standard does contain essential job functions that the physician may reference when making his or her recommendation, it is certainly advisable that the department provide the physician with its own position-specific job descriptions containing essential job functions that have been validated to the department's actual job requirements.

To ensure that current employees remain medically fit, it is recommended that they undergo regularly scheduled medical exams throughout the course of their employment. While some departments opt to provide medical exams on an annual basis, others set a schedule for exams based on the age of the fire fighter.

When developing a medical exam program, chief officers will also need to address the financial cost of providing the exams. However, these same chiefs can cite the benefits that healthy employees bring to the table and the potential of reduced medical costs for the employer when selling their program.

Employee Ownership

Even the best-designed health and fitness programs can fail if the employees do not support them. Employees must take ownership of their health issues. Remember the old saying—you can lead a horse to water but you cannot make him drink—well, you can purchase and install the best exercise equipment, but you cannot make the employee use it. Each employee must accept the fact that he or she is responsible for maintaining his or her own health and fitness levels. It is the job of the chief officer to establish a culture that instills personal ownership in health issues.

The best way to increase employee ownership in any wellness and fitness program is to involve them from the beginning and throughout the program. When employees feel involved and have input into the decision-making process they are more apt to support the end product. Chief officers would be advised to utilize the committee process when developing the components of the wellness program. In a unionized career department, the chief officer would be better served by asking the union to join a joint labor-management committee for this purpose. Buy-in from the union is important, especially when some of the desired wellness initiatives may need union approval through the negotiation process. Employee input can involve reviewing job descriptions for accuracy of tasks performed, recommending types of exercise equipment to be purchased, introducing healthy food choices in the fire station at meal times, concurrence with the medical examination schedule, and developing and testing a physical ability test.

Another area where employees can become involved, as well as remain active, is through the formation of a health and safety committee. Health and safety committees can be an important part of an overall wellness and fitness program. Chief officers can utilize the services of the committee to review modifications and make recommendations to the wellness-fitness program and, once established, compile data and evaluate the success of the program, review accident and injury reports, and make recommendations and review policies and procedures for safety issues. Committee members can become the best advocates for the wellness-fitness program to the other employees. Their testimonials are usually better received by the troops than are directives from the boss.

Exercise Time

When asking employees to commit to a comprehensive wellness-fitness program, the chief officer is asking them to make lifestyle changes. These changes can affect employees both on and off the job. As such, chief officers must be willing to commit to accepting change themselves in support of their participation FIGURE 5-3. The point in establishing a wellness-fitness program is to make improvements in the wellness and fitness of employees. If the organization insists on doing things the same way it has always done them, then it can expect the same results.

One area where disagreements can occur is when providing time while on duty for employee exercise and use of weight equipment. A chief officer might quickly jump to the opinion that if employees want to exercise while on duty they

Courtesy of Robert Klinoff.

FIGURE 5-3 Chief officers must walk the talk when asking department members to participate in a wellness-fitness program.

should be able to exercise during their evening downtime. The truth is that the department benefits as much as, or perhaps more than, the employees when those employees are physically fit for the task assigned. If one believes this, then coming to a compromise over workout times is beneficial to all concerned parties. If the employees are under contract, then workout time may become a subject of bargaining. A solution to the issue may require that a letter of understanding be developed or that the issue be discussed during the next negotiation process and included in the collective bargaining agreement.

Regardless of how an agreement is reached on workout times, the chief officer must understand that if the employees do not agree with the time they will probably decline to participate. If the chief officer is able and decides to make the workout mandatory, employees may show up at the required time and use the equipment. However, one cannot mandate effort, and effort is what determines success.

Rehabilitation of Employees

Employee rehabilitation can be required because the employee failed a medical examination, was unable to complete a mandatory physical ability test, or is returning from an injury leave. Whatever the case, the rehabilitation of employees should be of primary concern to the chief officer. It is important to show all employees that the wellness-fitness program is implemented to support the employees and keep them on the job; it is not punitive.

When employees are injured or unable to perform the duties of the job due to a medical concern, their absence can have a negative impact on the entire organization. Filling vacant shifts can increase personnel cost through the impact of additional overtime cost. On-the-job injuries have the added expense of workers' compensation payments, which can also lead to increases in workers' compensation insurance premiums. The injured employee might hold a

For example, if other local fire departments have sick day benefits that include 10 sick days annually and your department provides only five sick days annually, then the department's sick day allowance may be considered inadequate. However, when determining adequacy, it is important to consider the entire benefit package. Even though a benefit in one area may be considered inadequate, in relation to the entire package it might not be. For instance, in the preceding example the department offering only five sick days per year may offer a benefit of several personal days off and an additional week of vacation compared to other local communities and departments.

Chief Officer Tip

Equitable Compensation

Compensation and benefits are evaluated on their equity on two levels:

1. Internal equity is comparing the employee's compensation with that of other employee groups within the same organization.
2. External equity is comparing the employee's compensation with that of employees doing the same job in similar-sized departments in other similar municipalities.

When an employee identifies an inequitable situation, he or she will attempt to fix the inequity. This can result in the employee altering his or her job performance and even leaving the organization to find a more equitable situation.

To determine whether the organization is offering adequate benefits, three questions can be asked:

1. Is our organization recruiting the desired employees?
2. Is our organization able to retain our employees?
3. Are we getting the desired job performance from our employees?

Be mindful that the compensation and benefits are only one variable affecting the answers to these questions. Other variables need to be evaluated as well before making changes to the compensation and benefits package. If inadequate and inequitable areas are identified, the next step is to explore new potential benefits that will close the gap between what is provided and what is considered to be adequate and equitable.

Gather New Benefit Information

Gather information concerning proposed benefits or alternatives or enhancements to the present benefits. This can be accomplished at the administrative level and/or by a committee composed of knowledgeable and interested personnel. Soliciting feedback from all employees when making changes will ultimately create a more desirable benefits package. Using a survey instrument will provide the organization with information that can be easily navigated. Looking locally is a good place to start by inquiring about benefit information from other departments in the area. Fire department associations may also be a source of benefit information. A chief officer should also review any union proposals for increased or new benefits, including comparing these requests to benefits in comparable departments. Obviously, the providers of specific benefits such as health and life insurance should be contacted in regard to specific benefit packages and cost.

Create a Proposal

Finally, it is time to formulate a specific proposal to improve the benefits. A benefit proposal should include the following sections:

- An explanation of the current or new benefit
- The current benefit's shortcomings or a new benefit's advantages
- A proposal for providing a new benefit or for improving and enhancing an existing benefit
- The justification and rationale for the new benefit or for proposed changes and enhancements of a current benefit
- The cost(s) of the proposed benefit changes
- Identification of any new or reallocated revenue stream to fund the new benefit package

When proposing a new benefit or changes to existing benefits, it is a good practice to provide a reasonable number of options for providing the benefit. Doing so provides the decision makers with a better perspective about what was researched prior to the proposal and helps provide perspective on the particular proposed benefit. This can be done by proposing a good, better, and best option. A good example is health insurance benefits. There are many options for coverage, deductibles, administration, and so on. In the process of finding options that may suit the intended needs, there are many different combinations for coverage and deductibles, all of which require different levels of funding.

Additional benefits are limited only by creativity, the willingness of the organization to provide the benefits, cost, and fairness and equity for all members of the department.

Personnel Management

A fire department's personnel are its most important asset. Good employee management relations are essential to a department's ability to provide the best possible services to the public. Management has obligations to its employees, and employees have obligations to management. These obligations are intertwined with the responsibility of providing the best possible services to the citizens they serve. Providing the best possible service takes effort and the cooperation of management and employees. A positive and participative management program requires an organizational environment that helps motivate its members to do their best FIGURE 5-8. When properly motivated, individuals of any organization are able to perform at a high level. Having leaders who are able to create an environment that helps people be motivated professionals is the key to any

VOICES
OF EXPERIENCE

With the ever-changing cultures in today's fire service, what was once an era when fire department leaders felt it was solely their job to design the next new fire apparatus or the next best training program, or even to handle the many issues around the firehouse, has turned into anything but a one-person show. More than ever, in today's culture, fire fighters are looking for ways to be heard.

I understand this philosophy could be viewed by some as a symbol of weakness and indecisiveness; however, I can promise this is not the case. It is vitally imperative that department leaders look beyond themselves for the next best idea or help in solving problems to any given issue within their own departments. If they do so, perhaps they may just find that solution to a problem that has held their department back for so long. As chiefs, we all have a sense of pride and belief that what we have created or done within our own departments is the best and is what works for us, but is it really? I think it is time for department leaders to examine their own departments and determine whether a cultural adjustment is needed, both at the top and throughout their organization.

It's time to step up and make the change toward empowering people to get involved.

Let me take a moment to ask a few simple questions: When was the last time you involved your staff in making key decisions related to the purchasing of new equipment? Did you involve your staff in the design of the last apparatus? How about this: When was the last time you called a meeting with union officials to discuss any concerns related to the operations of your department? The common thread to all the above questions is that these situations provide an opportunity for staff to have a voice and provide some ownership into key decisions being made. Now, I'm not saying to simply go through the motions; you must encourage your staff to get involved, you have to demonstrate your willingness to provide the stage and to seriously take note of their suggestions. Taking a quote from the movie Field of Dreams "build a field and they will come." I say, "build a stage and they will come."

I have personally witnessed the drive and determination by two union groups, one part-time and one full-time, within my own department, who worked together to offer solutions to address an operational issue dealing with shift filling procedures. Their solutions not only met the needs of the two union groups, but what they felt were the needs of management as well. It was very early on in the process when the two groups felt that this type of collaborative effort was too powerful to utilize in solving a single issue.

With the addition of a management team member, a more formal group was formed into what became known as the Joint Relations Committee (JRC). The purpose of this committee was to discuss delicate issues between bargaining groups and management and issues between the separate bargaining units as well. This was a big step in defining how the tough issues would be dealt with moving forward. It was evident how important it was to the labor groups to have a voice in the outcome of certain situations.

VOICES OF EXPERIENCE

(Continued)

Although the first attempt by the committee failed to produce positive results on the shift filling issue, it did lay the framework for how labor and management could work together for a common goal. For example, both labor groups and management later formed a Health and Wellness committee whose goal was to see to it that fire fighters were in the best possible shape to perform their jobs. The committee developed a program that consisted of mandatory health physicals provided by the employer per NFPA 1582 standards. Also, it included designated optional workout times for fire fighters while on duty. In addition, the program required that all fire fighters complete a mandatory annual physical ability test as a condition of employment. This Health and Wellness program received honors and was awarded the Senator Paul Sarbanes Award at the CFSI dinner held Washington, DC in 2008. This level of achievement would not have been possible had it not been for the collaborative efforts by all work groups.

As a chief, let me be the first to say that it's a work in progress. It takes a tremendous amount of effort each and every day to maintain a positive cultural environment enriched with ideas and progressive thinking throughout all levels of the organization. There are days when I ask myself "is this all worth it?" Of course, the answer is yes. It is our jobs as leaders in the fire service to provide a model for the level of professionalism we expect from our future leaders. It's not time to give up; it's time to step up and make the change toward empowering people to get involved. I encourage all fire service leaders to take the time to examine their own department's culture, to make adjustments where necessary and, if needed, build the stage.

Bob Gagnon, BBA
Fire Chief
City of Norton Shores Fire Department
Norton Shores, Michigan

FIGURE 5-8 The best organizational environment is one that motivates its members to do their best.

good organization. There are many sources of information on the topic of employee motivation; studies, books, and articles on the subject are widely available.

The relationship the employee has with the organization is invaluable. Employee engagement includes the recruitment of new employees, retention of current employees, conflict management, the discipline process, and the performance appraisal process. It is not good enough for the organization to do just one of these well; the organization needs to be successful from the first interaction with the employee and must continue building on that relationship throughout the employee's career. Just as the chief officer is evaluating a potential new hire, that same new hire is evaluating the organization. The better the organizational fit between the employee and organization, the more valuable the employee can become and the more efficient the organization will be in meeting the needs of its customers.

Evaluating Management–Employee Relations

Evaluating management–employee relations is the first step in determining how well all the people in the department work together to provide the best possible services to the department's external customers (the citizenry) and internal customers (department personnel). The concept of internal and external customers is important to understand in relation to a department's ability to deliver high-quality services. If internal customers operate in an environment that does not motivate them to be highly professional, then chances are that the personnel delivering the service to the citizenry are not operating at an acceptable level. Generally speaking, the better the management–employee environment, the better the employees, and thus the better the service to the community.

The key to determining management–employee relations is to examine the factors that produce motivated employees and those that do not. From this perspective, specific motivating or nonmotivating factors can be identified. Identifying these factors can be done in a variety of ways such as surveys, exit interviews, discussion groups, officer retreats, outside consultants, and simply by asking personnel to speak their minds honestly about what the department is or is not doing to help them be motivated.

A positive and participative management–employee program can then be developed to address and correct the nonmotivational factors that have been identified. Developing a process involves the basic problem-solving model in which the identified problems are prioritized, options for fixing them are identified, the best option to solve the problem is chosen, the solution is implemented, and the effectiveness of making any necessary change to accomplish the desired effect is evaluated.

Communication

There is no disputing the fact that chief officers who fail to master the communication process will have more struggles with employee relations than those who are able to communicate well. Nonemergency communication skills are also important to the chief officer. While the ability to speak publicly is certainly important, many chief officers underestimate the importance of being able to write legibly, with clear thoughts. Achieving proficiency in writing skills enables the chief officer to sell ideas, manage programs, and interact positively with subordinates, peers, and supervisors.

The inability to communicate effectively and/or clearly has created many internal problems for the chief officer. The level of employee morale is often determined by the chief officer's ability to communicate effectively with the troops. Keeping employees and supervisors informed should be a priority for any chief officer and begins with mastering the many forms of communication available. Additionally, it is well documented that the failure to communicate effectively at an incident scene places the safety of fire fighters at risk. Many near misses, fire fighter injuries, and line-of-duty deaths have some type of communication failure as a contributing factor. Chief officers must learn how to communicate effectively under the pressures of an ever-changing emergency scene. Fully understanding the implementation of the incident command system and achieving proficiency in communications technologies can greatly assist the chief officer with improved communications. (For more information on communication, see the "Communications" chapter.)

Hiring and Recruiting

Hiring and recruiting are critical to the human resources of any department, whether career or volunteer. Proper hiring in career departments and member appointment in volunteer departments are very important because the policy and procedures must not only comply with all applicable legal and ethical issues, but they must also ensure that the persons allowed to be members of the department are well qualified and have the potential to become good members of the department.

Recruiting potential candidates for fire department membership is the first step in the hiring process. The goal of a recruiting process is to find potential candidates who fit the job description of the position to be filled. These potential candidates can be found in the general population of the community and beyond. Before the positions are advertised for hiring, a job description statement and the basic predetermined qualifications should be stated, such as minimum age, driver's license requirements, citizenship documentation, minimum education, physical ability, and so on. Also, basic information about the application and hiring process should be included. This information can then be disseminated in such a way that all potential candidates have the same access to the recruitment advertisement. In some cases, recruitment efforts are targeted to specific groups to ensure that department diversity goals are being addressed.

There are many ways to advertise the open position(s) the department intends to fill, including media such as local newspapers, websites, radio, direct mail to organizations such as community groups and local education organizations, and speaking presentations to individual organizations. The important thing to remember is that the advertising process and subsequent parts of the hiring process must be compliant with all legal and ethical requirements in regard to recruiting and hiring.

In the hiring process, determining the qualification and potential of prospective members can require many components. These components fall into several categories of assessment; for example, three of these components can be described as domains of learning—cognitive, psychomotor, and affective (e.g., cognitive ability evaluation via written testing, psychomotor evaluation via physical fitness testing, and affective evaluation via oral interview). Assessment centers, interviews, and written tests may all be used to evaluate the candidate. Other component categories include a background assessment and an overall health assessment. Remember to follow federal employment guidelines because some evaluation components (e.g., health examinations) can be done only after a conditional offer of employment is given.

Life Safety Initiatives

6. Develop and implement national medical and physical fitness standards that are equally applicable to all fire fighters, based on the duties they are expected to perform.

NFPA standards such as NFPA 1001, *Standard for Fire Fighter Professional Qualifications*, and NFPA 1582 can be helpful references in determining the components of the hiring process. Local, state, and national fire service organizations may also provide basic information concerning hiring and recruiting. The IAFC and IAFF jointly developed a Candidate Physical Ability Test (CPAT) for use in evaluating a potential candidate's ability to perform the essential job tasks of a fire fighter.

A very important but often overlooked factor for a potential new hire is organizational fit. A better organizational fit can result in the hiring of a more motivated employee. Those who "fit" the organization tend to get along with other employees, causing fewer personnel issues. They also are more likely to support organizational goals and objectives, work hard to advance, and require less supervision while remaining motivated in their job. Ultimately those employees who display organizational fit qualities are simply happier in their chosen profession, which will usually carry over to the service they provide to the customers they serve. To ensure the best organizational fit with a new fire fighter or officer, the process starts before the candidate even applies to the department. It is the responsibility of the department to market itself to the community, and perhaps more important, to be able to target potential candidates. Even before a potential candidate gets an application, the department can require individuals to view department literature or multimedia presentations. This might include testimonial-style interviews with current employees explaining the challenges they face or why they love the job. A fire fighter might speak about the challenge of missing important milestones in his or her child's life because of shift work or the satisfaction of making a difference in a stranger's life. It is important to provide a real picture of the department, both benefits and challenges.

As the candidate advances through the process and gets closer to receiving a job offer from the department, the department should step up the organizational education. The department should strongly encourage or require a ride-along type program so the potential employee can feel the organization's culture and experience some of the professional expectations. It is imperative that all legal guidelines are followed with any ride-along program. The ride-along can be positioned in the process in such a manner that it does not overwhelm department resources with a large number of candidates. One example would be having a ride-along completed prior to an interview. This would allow the opportunity for an organization-specific question to be used in the interview. The specific timing of a ride-along program is dependent on numerous organizational variables.

Whichever processes are used, the overall goal in hiring and recruiting in volunteer and career departments is to ensure that the department's most important resource, its personnel, are recruited and hired in such a way that the new member has a good potential of becoming a valuable part of the department.

Employee Assistance Programs

Fire service organizations are not much different from any other organization in that employees' personal issues can affect job performance. These issues can be relatively minor in nature or rise to become serious situations. Personal problems that affect job performance can include personality conflicts among peers or supervisors, problems outside of work (either personal, financial, or relationship related), addictions of many varieties (physical as well as mental), and physiological disorders. Remembering that department personnel are the department's most important asset, a department must be in a position to offer help and guidance to its members when the need arises. Providing an EAP is a necessary component of good human resource management and support.

Life Safety Initiatives

13. Fire fighters and their families must have access to counseling and psychological support.

EAPs can take many forms. Some options include professional EAPs delivered by outside organizations and fire department chaplaincy programs. EAPs are provided by professionals trained in mental and psychological health issues that may emerge in any person. These professional counselors are fully trained and certified and are capable of helping solve the personal problems of employees. They may be able to provide treatment in specific areas or may refer a person to the appropriate treatment. They can also provide programs for good mental health and overall wellness. In some cases, employees are mandated to participate in an EAP as the result of a discipline issue. Further information on EAPs can be obtained from the Employee Assistance Professionals Association website.

Local professionals with counseling practices may offer a variety of personal and professional counseling services. Internal assistance can also be made available, such as peer conflict resolution teams, mentoring, and professional development programs FIGURE 5-9.

EAPs that are already in place can be evaluated to determine whether the program is having its desired effect. Critical items to evaluate are the program's competence with the job responsibilities, accessibility, and the members' confidence in the program. Employees may seek access to an EAP for any number of reasons. While EAP counselors are used to dealing with the many personal issues that employees encounter, it may be advisable to determine the EAP counselor's experience and ability to deal with the types of traumatic issues handled by fire service personnel. A department may experience problems when personnel who are referred to an EAP cannot relate to the person they go see for treatment because the counselor has no training or experience with people who witness first hand death and destruction, sometimes on a daily basis.

FIGURE 5-9 As part of an EAP, counselors can provide treatment in many specific areas or can refer a person to the appropriate treatment.

Any good assistance program must be accessible to all personnel, and the personnel must feel confident in the program's ability to help solve a given issue. Most important, the program must ensure confidentiality. If employees feel that a program will not be helpful or confidentiality can be compromised, they simply will not take advantage of the program.

Chief Officer Tip

Evaluating an Employee Assistance Program

One way of assessing EAPs is to survey personnel to find out what they think about a current program. Would they seriously consider using the program if needed? Have they ever used the program? Has the program helped them? Do they have any suggestions for improving the program?

Another evaluation method is to conduct a confidential review of past problems that were referred to the program. This should be done at the appropriate administrative level (i.e., the person who has the highest authority and responsibility in the administration of the assistance program). The review must be completed with due regard for any legal and policy or contractual guidelines. Using the appraisal information, any deficiencies or problems can be identified and improvement can be initiated. Maintaining an effective, accessible, and confidential EAP is an important part of the responsibilities of a human resources officer.

Conflict Management

Conflict management is a valuable and required skill for chief officers. Chief officers often find themselves in circumstances where good conflict management skills are needed. Conflicts can arise in all ranks within a department and sometimes rise to the chief officer level. As with all fire department operations, it is important to have a well-defined chain of command for resolving conflicts. When conflicts do arise, they should be mitigated at the lowest level possible within the department; the chain of command within the department should be followed. When a conflict is presented to a chief officer, the first question to be answered is, "Has every appropriate effort been made to resolve the conflict prior to the intervention of the chief officer?" If not, the chief officer should refer the conflict to the appropriate person within the chain of command.

In some cases, conflicts may be forwarded directly to the chief of the department due to the serious nature of the complaint. One example might be a conflict between two or more employees that involves a form of harassment or violence in the workplace. In addition to rising directly to the chief officer, these issues may also be immediately referred to the organization's human resource officer or personnel officer. The chief officer may have to provide support and encouragement to the person responsible for resolving the conflict and follow up to see that the issues have been addressed.

If the conflict does rise to the chief officer level, all applicable facts of the situation should be gathered and analyzed to determine the right course of action. This involves keeping an open, objective mind-set that will help in determining the basic issues involved. Remaining objective is especially important when the conflict has embedded personality issues of those involved. Sometimes a conflict may be a result of underlying issues that need to be addressed. Determining the best course of action is dictated by the facts of the situation and consideration of any legal, contractual, policy, or procedural requirements that may apply to the situation. Also, the course of action in regard to resolving the conflict should be consistent with past good practices and good leadership ethics.

Chief Officer Tip

Conflict Resolution

The chief officer must avoid the initial desire to jump in and take control of a conflict situation when the supervising officer has not yet had an opportunity to address the issue. Chief officers must let their chain of command work by referring the employee with the problem back to his or her supervising officer for resolution to the problem as the first step in the process. The chief officer may have to provide support and encouragement to the person responsible for resolving the conflict and then follow up to ensure that the issues have been addressed.

Of course, there is an exception to most rules, and conflict resolution is no different. There are some issues that may be first brought to the chief's attention by an employee that should be addressed by the chief officer. Examples may include issues dealing with sexual harassment or a hostile workplace complaint. Given the severity of these issues and the potential liability for them, these issues are normally dealt with directly through the chief's office with assistance of the organization's human resource department and possibly the advice of legal counsel.

Discipline

As a chief officer, the roles and responsibilities related to the disciplinary process are different from those of a company officer. First, a thorough review of the disciplinary process needs to be conducted, particularly if there is not a clear process in place. One of the first steps is making sure the organization's policies and procedures are conveyed to the employees and that the employees understand the policies and procedures. A common practice is having employees sign that they have read and understood an Employee Handbook or similar organizational publication. Additionally, it is necessary for chief officers to model behavior that they expect from their employees. When the employees see that the rules apply to everyone, including the chief, there is less opportunity for individual interpretation or selective rule following. The chief officer needs to be consistent and fair in his or her actions and decisions as they relate to discipline. A lack of consistency and fairness will create confusion and frustration in the organization.

Discipline for infractions should be clearly defined in the policies and procedures. Therefore, if an infraction occurs, the resulting action from the organization should not be a surprise to the employee. This degree of predictability helps employees understand the consequences of their actions before they err. This process can help establish self-discipline in the employees. The goal of the discipline process is to modify an employee's behavior when that behavior is not meeting organizational expectations and/or is not in accordance with organizational policies and procedures. The discipline process is successful when an employee changes his or her behavior to meet expectations, not when discipline leads to termination.

Discipline should be used only when all other means of achieving subordinate compliance with rules and regulations have been exhausted. Employee counseling and proper use of the employee evaluation process may assist in mitigating the negative behaviors that, left unchecked, may ultimately result in the need to administer discipline.

Identification of a problem will lead a supervisor either to handle the problem at his or her level or to recommend some type of escalation in the chain of command. There must be a clear policy in place that determines which infractions should be escalated in the chain of command. A good example of a policy violation that would trigger escalation would be a suspected substance abuse issue. Another trigger for escalation in the organization would be a reoccurring infraction and/or behavioral issue. If a supervisor feels an infraction is significant enough, he or she always has the ability to escalate that issue to the next level.

If the supervisor is going to handle the problem at his or her level, he or she can follow some simple guidelines to make sure the organization is meeting the employee's needs:

- Identify the policy or procedure that was violated
- Investigate to make sure information is correct
- Talk with the employee about the infraction
- Understand the entire situation
- Decide on behavior change
- Get commitment from the employee
- Document the meeting with the employee and set a follow-up schedule

When handling an issue with an employee, the chain of command must be kept informed about the situation, actions being taken, and the results. Department policy will determine specific guidelines for supervisors in handling discipline issues. It is also important to consider any compliance issues with collective bargaining agreements; other union contracts; and federal, state, and local laws.

Supervisors should document all employee performance and behavioral issues, both the positive and negative. Employees may be concerned that something is being noted in their file, but this documentation is necessary. If there is no documentation, it did not happen. The documentation will also assist the supervisor in completing future performance evaluations. To add transparency to an already intimidating process, the supervisor and employee can initial any documents they have read. This initialing, however, does not necessarily mean they agree with the material of the document.

To make the disciplinary process less intimidating for both the supervisor and the employee, consider assigning a responsible party (ideally someone from the executive staff) to receive and transmit information related to the process. Having a clear destination for the information yields better communication relating to disciplinary issues.

> **Chief Officer Tip**
>
> **Investigating an Infraction**
>
> When investigating a potential infraction, seek answers using the basic questions of journalism:
>
> - Who
> - What
> - When
> - Where
> - Why
> - How
>
> How this information is put together will be organization-specific.

Investigation into an infraction should be focused on facts. The judgments and opinions of the investigator should be omitted during this process. One of the most important things to keep in mind during the investigation is proper and thorough documentation. Once the investigation is complete, the chief officer should review all of the information before reaching a decision. Some organizations might benefit from having a panel review the recommendation of the chief officer. Whatever decision is reached, the discipline measures should reflect the seriousness of the offense and the employee's overall record of service. It is also important that the chief officer follows through with the discipline. A supervisor who has to handle a discipline situation will appreciate the support of the chief officers and the organization.

Progressive Discipline

The concept of progressive discipline dictates that discipline severity escalates with each infraction over a specific time period or similar infractions over a long time period. The organization is responsible for defining the time period for a progressive discipline procedure. General guidelines for the progressive discipline process can be defined, but it may also be necessary for organizations to define specific disciplinary actions for specific infractions. A chief officer must take the organizational culture into consideration when determining the progressive discipline process. Steps will generally follow a particular order, but for serious and significant incidents steps might be skipped. For example, it may be necessary or desired to terminate an employee caught stealing on the first offense. It is imperative to follow through with the specific discipline; otherwise discipline will never take hold in an organization.

An example of a progressive discipline process is as follows:

1. Counseling session
2. Verbal warning with documentation
3. Written warning and possibly other disciplinary measures
4. Second written warning, including other discipline and the stipulation that failure to correct the behavior will result in additional disciplinary action up to and including termination
5. Suspension, usually 1–30 days
6. Last chance agreement
7. Termination

At each step in the process, the employee should receive detailed feedback on what actions he or she needs to take to correct the behavior and what discipline would be enacted if the employee becomes a repeat offender. The goals of the disciplinary process are to correct the behavior and increase employee performance with the least amount of discipline necessary.

Predisciplinary Hearing

In most cases, a predisciplinary conference or hearing (also referred to as a *Loudermill hearing*, based on *Cleveland Board of Education v. Loudermill*) must be conducted before a suspension, termination, or involuntary demotion can be invoked. The degree of investigative effort and the opportunities for employee response before these punishments are issued are set higher than for less severe levels of discipline. This step ensures the employee has been given due process. (See the "Legal Issues" chapter for more information on Loudermill hearings.)

Last Chance Agreement

A last chance agreement can be used immediately for significant discipline issues or in the final stages of progressive discipline before termination. In the last chance agreement, the employee agrees to make a behavior change and meet the organization's expectations. This agreement outlines that the employee understands that this is his or her last chance to make the necessary changes and that if he or she does not meet these expectations or make changes in behavior in a specified period, his or her employment will be terminated. A specific plan for employee improvement can be outlined in either this document or a separate document but should be referenced.

Garrity Rights

Garrity rights are protection for the public employee as it relates to criminal prosecution. These rights are derived from the Fifth Amendment, noting that an individual has the right not to incriminate himself or herself. If the public employee is compelled to answer questions during the course of the employer's investigation, those statements cannot be used against the employee for criminal prosecution. It is important to seek legal advice when this situation develops. These rights were established based on *Garrity v. New Jersey*, in which the court held that a police officer compelled to make a statement or be fired, and then subsequently prosecuted criminally based on that statement, was compelled to incriminate himself in violation of his Fifth Amendment rights. (See the "Legal Issues" chapter for more information on Garrity rights.)

Chief Officer Tip

Negligent Hiring and Retention

Employers can expose themselves to negligent hiring and negligent retention based on a few simple considerations: Did the employer know (or should they have known) the employee was dangerous, unfit, or posed a threat to others?

Employee Evaluations

When properly implemented, employee evaluations can be a useful tool in improving employee performance. Employee evaluations are best completed by the employee's immediate supervisor because that officer is typically in the best position to observe firsthand the employee's attitude, behavior, and productivity. It is also acceptable to have multiple officers contribute to the evaluation process, especially when employees are not assigned to shift work (e.g., part-time or volunteer employees). When multiple officers provide input on evaluations it is still best to have just one officer deliver the evaluation to avoid the impression of ganging up on the employee. In some cases, employees are asked to self-evaluate first and then submit their completed forms to their supervisor for review and additional input by the supervisor.

Evaluations should be completed on an annual basis and are often scheduled for completion around the employee's anniversary date. For new employees, the evaluation period may be shortened (e.g., six months) to provide early feedback to the employee during the probationary period. The evaluation becomes part of the employee's permanent personnel record and may then be referenced during future promotion opportunities. Because evaluations are a part of the personnel record, they should be completed in an open and honest manner. More than one department has found itself in a difficult situation when trying to formally discipline an employee for bad behavior only to have that same employee refer to multiple positive evaluations on file and ask, "Why have I not been told of these issues before?" With incomplete records, an employee may challenge a disciplinary action, potentially placing the employer in a compromised position if legal action is involved.

The evaluation process should be approached from a positive frame of mind and not be used as discipline. The evaluation is a supervisor's chance to point out both positive employee behavior as well as areas needing improvement. In cases where employee performance is not up to department standards, the evaluation interview should focus on discussions that explore ways to improve the negative or below-standard performance. An emphasis should be placed on correcting the deficiencies noted with a supervisor's commitment to assist the employee in any way possible. Improving an employee's performance may require that a supervisor and employee agree on an improvement plan with follow-up reviews scheduled throughout the year.

There are many ways to construct the evaluation instrument. In some cases a simple numeric system is used to rate an employee's performance on a scale of 1–10. Other systems use descriptors such as "needing improvement," "meeting expectations," or "exceeding expectations." Regardless of the system being used, it is important that the employer provide the supervisor with training on how to complete the evaluation process.

One of the most important aspects of any evaluation program is consistency in evaluating employees. This is especially true when multiple supervisors are involved with different employees. It is unfair for one supervisor to rate his or her employee group consistently higher than another supervisor's group, especially when those evaluations are used as part of an employee promotion process. To maintain accuracy, supervisors should be instructed to keep supervisory notes on employee performance throughout the year to avoid having to rely on memory when completing the evaluation instrument. Citing specific examples of both negative and exceptional behavior during the evaluation is a good way of validating the evaluation given.

Employee evaluations can also be used to mentor the exceptional employee. In these cases the evaluation interview can focus on setting a career path and discussing opportunities for expanding the employee's horizons. Additionally, the supervisor might be able to suggest how the employee can further his or her education, training, and experience in preparation for promotion opportunities.

Development of Human Resources

Professional Development

Chief officers should be involved in the professional development of their staff. Professional development starts with assessment of an individual's current professional status; which KSAs a person has at present; and what areas need to be improved. The next step is to identify career goals and objectives, as well as the KSAs necessary to attain those goals and objectives. A career goal is a statement of where a person wants to be in the department, and the objectives are the necessary steps to get there. A goal for a new member may be to become a company officer within five years; the objectives are the steps necessary over the next five years to attain that goal.

After professional goals and objectives are identified, the next step is to develop a well-thought-out plan to complete the objectives and attain the goal. The plan should identify each step to be taken in a specific time frame. The plan should be monitored and adjusted as circumstances change. Implementing the plan is a function of the opportunities available and the individual's motivation and desire to complete the plan and overcome any problems along the way. This process may be repeated several times during the course of a career depending on personal or organizational circumstances.

There are many steps in the professional development process, most of which involve a training or education component. Training and education opportunities in professional development can include all domains of learning—cognitive (thought process/comprehension), psychomotor (perception/motor skills), and affective (attitude/valuing). Management and leadership training and educational opportunities are necessary in preparing members who have professional goals to become line, staff, or chief officers. Additionally, current chief officers are invaluable as mentors to new officers and in helping the department transition to new leadership. (See the "Personal and Professional Development" chapter for more information on professional development.)

Encouraging Participation

Encouraging members to participate in professional development is part of creating a culture of continuous improvement. If personnel do not participate in professional development, how can they and the department improve? Members can be encouraged to participate by offering and supporting professional development programs that are relevant, accessible, and as convenient as possible. This also includes fair and equitable access to high-quality training and educational opportunities. It is very important for the organization to maintain a structure that provides a variety of opportunities that allow increased authority and responsibility. Discussion of professional development options should be part of the employee evaluation process (discussed earlier in this chapter). These opportunities should not be limited to rank promotion. Other prospects for increased authority and responsibility should be offered in all areas of the organization. Opportunities can be offered in any area of operation from management information system coordinator, to specialty team leader, to public education safety specialist. Every department needs personnel who have specialized talents; a professional development program is the best way to equip personnel for these important roles within a department **FIGURE 5-10**.

Training Program Management

Training and education are important components of any professional development program. It is understood that all new hires will complete some type of entry-level training program or academy, but what happens after that? While probably not responsible for the direct delivery of training, a chief officer is responsible for ensuring that a comprehensive, department-wide training program is in place that properly prepares his or her fire fighters to provide safe and effective response services.

The chief officer plays an important role in establishing the importance of training within his or her department. Training programs are established to ensure that fire fighters are prepared to meet the response needs of the community. While the subjects and competencies can vary among departments, there are some similar components that should be considered when developing a department training program. The components of a training program should address the following areas:

- Establishing a training policy
- Identifying a training officer
- Appointing a training committee
- Establishing a training schedule
- Acquiring training equipment
- Training instructors
- Monitoring results

The chief officer sets the tone, and ultimately the importance, of the training program by establishing a training policy. Training policies are designed to identify the amount and type of training as well as the discipline for noncompliance with the established requirements. Although training policies can be developed solely by the fire chief, it is advisable to involve a cross section of the department. Established training requirements will be better received if the individuals being held to the standard are allowed to provide input.

The amount or frequency of training will depend on the type of fire department structure. Career departments may be able to justify training on a daily basis because personnel are available on predetermined shifts, whereas volunteer departments may meet only twice a month, with training sessions competing with fire fighters' personal family time and full-time employment.

Training topics need to be determined after a review of state and federal requirements. Job descriptions and other requirements will influence the selection of training topics. Additionally, a community risk assessment performed by the chief officer may help determine the priority of both type and number of training topics needed.

To assist in managing the training program, the chief officer may assign someone to the position of training officer. This position may be filled differently in each department. In some cases, the training officer duties are simply assigned to one of the existing line officers. The position may also rotate among several line officers on a predetermined schedule. In Michigan, Public Act 291 mandates the appointment of a department training officer, but there is no requirement that that individual hold a ranked officer position. The training officer serves as the official spokesperson for the training program and is charged with overseeing the development, presentation, tracking, and coordination of the training program.

One sometimes fatal mistake that chief officers make is placing all of the responsibilities for the department training program on the back of a single officer. Even an energetic individual can get burned out when required to provide the entire department's training needs. A better approach might be to appoint a training committee with the training officer serving as the committee chair. The use of a training committee allows multiple personnel with individualized areas of expertise to be involved in the training process. This spreads out the workload, not only helping to prevent burnout but also diversifying the delivery of training topics.

One of the first tasks to be completed by the training committee might be to establish an annual training schedule. Advanced notification of the training requirements and scheduled events will allow personnel to plan in advance

FIGURE 5-10 Professional development programs can be used to train members in specialized skills.

Reductions in municipal revenue have taken their toll on your department as well as surrounding departments. Staff reductions have been implemented and apparatus purchases delayed or even cancelled, while at the same time requests for service have increased. Your chief elected official has recently returned from a state conference where presenters were touting the financial benefit of shared services. He has asked that you examine solutions to your response needs in the context of greater cooperation between your department and those surrounding you by gathering information on the following issues.

1. What is the basic difference between implementation of a mutual aid agreement versus an automatic aid agreement?
2. How would the implementation of an aid agreement affect other areas of the department's operation?
3. What are the lasting benefits of fire department involvement with an intergovernmental activity?
4. How does the fire department avoid creating a polarizing political issue when engaging with other agencies?

Introduction

Fire departments that operate as if they are independent entities are not taking advantage of the many beneficial relationships available throughout the community. In these cases, the chief officers responsible for leading these organizations fail to recognize that their departments are an integral part of the community in general. Chief officers today must understand that community support is elemental in developing a successful fire service agency. Without community support, many departments struggle to obtain the necessary financial resources that drive every aspect of their service.

In most small- to medium-sized government communities, the city council, village council, fire district board of directors, or township board of trustees functions as an oversight committee. In medium- to large-sized cities, the fire department is generally an arm of local government. There may be someone on the local council or committee charged with being a liaison to the fire department. In some large cities, there is a politically appointed civilian fire director or commissioner. This individual reports to the governing body and is charged with developing and implementing the fire protection policy of the community.

Fire department operations are placed under the command of a uniformed fire chief or other level of chief officer. A fire department oversight committee can take on various members from the city or town councils; in most cases they maintain internal representation from existing members as opposed to private—for example, police, public works, streets department, village administration, and private citizens as volunteers. The interaction between the fire department and the local government is an important element in the successful delivery of services. The relationship between the local government and the fire department is critical to the development of a proper service delivery system. A lack of oversight or an absence of interaction can allow a fire department to drift aimlessly, losing focus and direction. This interaction is especially important in times of economic stress; chief officers must make their departments as visible as possible, in a positive light, to facilitate funding.

Chief fire officers must be prepared to interact with a wide range of local, county, state, and federal agencies and must identify the key players at the local, state, and federal levels and in specialty groups with which the chief officer must interact.

Fire Officer III

Roles of the Chief Officer

Regardless of whether a fire department includes career, volunteer, or a combination of career and volunteer fire fighters, the chief officers charged with running the agency must remember that the citizens within the community are the true focus of fire department operations. Chief officers who are inwardly focused risk alienating the members of the community at large, ultimately losing their support and encouragement. Without a community's support, the chief officer will find it difficult to find opportunities for additional funding, revenue enhancement through referenda, and various grants.

Interpersonal skills are a critical element within an officer's tool kit and help him or her perform better. In this way, he or she is better able to further his or her career and provide a better level of service to the community.

Communications skills assist the chief officer in making the public aware of issues important to the fire department.

Chief officers should continually evaluate and develop their abilities in the areas of written and oral communications. Opportunities to expand and improve the chief officer's communications skills should be explored. Examples could include a community college course in persuasive writing or a course in public speaking. (See the "Personal and Professional Development" chapter.) The public is constantly bombarded with media messages, and the fire department is but one of the many groups competing for the public's attention FIGURE 6-1. This can be a critical element within each of the areas in the chief fire officer's performance tool kit.

There are a number of trends in the world of government operations that the chief fire officer must understand to avoid appearing out of date or out of tune with the rest of the community.

Community Support of Residents

In any community, the residents are the key factor in determining the type of fire protection and EMS care received. Largely, these services are supplied through taxation and property values, in which residents have input. Many small communities rely on support through fundraisers for the volunteer services that protect their community. Other large metropolitan areas rely on taxation to fund essential services. Still, the community, and ultimately the fire department, needs the support and input of its residents to determine the type of services or needs they expect. This expectation brings forth the residential support.

A fire department must interact with the community to gain insight into the community's needs. This can be accomplished by developing a citizens' advisory board (CAB), which can be dovetailed into the department's public education program and can improve community relations. Neighborhood community programs like the Citizens Fire Academy, Feel the Heat, CPR, CERT, Fire Corps, and babysitting classes promote community involvement. From these programs, the fire department can track and collect specific data that will show the effectiveness of the educational programs being offered. It is important to include a review of the community within the department's various planning efforts. (See the "Working in the Community" chapter.) One such program is smoke alarm education updates. The use of carbon monoxide detectors is a key to bringing the community together on awareness issues, especially after a local or well-publicized incident that has the community asking how to improve protection. Working with civic organizations can defray the equipment costs because these organizations often donate smoke alarms and carbon monoxide detectors to the programs.

FIGURE 6-1 Media and public awareness skills help the chief officer inform the community about issues important to the fire department.

There should also be a community element within the department's strategic, comprehensive, and master planning operations, which are conducted in coordination with local government. (See the "Strategic Planning" chapter.)

CPR Programs

One of the ways a fire department can build a strong bridge into its community is to offer a series of CPR programs to their citizens FIGURE 6-2. In this way, the department will be able to demonstrate a true concern for the well-being of the residents of its community. CPR programs can be developed to target segments of the community including seniors, school-age children, and adults. And while the public has long been engaged

> **Chief Officer Tip**
>
> **Going Green**
>
> There are a number of tasks a fire department can undertake to be considered a part of a community effort at going green. The use of solar panels or windmills to generate electricity could result in a reduction in the fire department's operational budget. In addition, the switch to fluorescent or LED lights can result in lower electrical costs and a reduced budget expense for lightbulbs. The use of biodiesel and E-85 fuels, where possible, can be put forward as an example of a fire department's concern for the environment, as can using fewer staff vehicles and more fuel-efficient cars. Finally, simple tasks such as recycling paper and plastics help confirm environmental awareness. Although making environmentally sound improvements may be expensive at the outset, in the long term, it can result in savings—not only financial savings, but also environmental and community relation savings.
>
> The chief officer may even form a "green" alliance with local government and private civic organizations. It is important for chief fire officers to realize the importance of thinking globally and acting locally. This is an area where chief officers need to think outside the box. Only a generation ago, many of the most complex and far-reaching environmental and socioeconomic issues were discussed only at the national and international levels. Today, local leaders should embrace and urge action on issues such as energy independence, natural resource conservation, and public health. They must recognize the opportunity to influence these issues. Leaders should consider it their duty to act, especially because the impacts of such problems are often felt first at the local level.

in CPR programs, they see fire service personnel as experts in this field. With CPR practices constantly under review, it is beneficial that the community continues to consider the fire department as an instructional resource. Since the emergence of hands-only CPR, more citizens may consider performing CPR on a stranger. The fire service is well positioned to provide such instruction to the community.

Feel the Heat

Another way to garner support within the halls of government is to conduct familiarization programs for members of a governing body and their aides. In Ohio, the state fire academy conducts a program known as Feel the Heat, and IAFF supports a program called Fire Ops 101. Both programs are designed to provide elected officials with an opportunity to understand what their fire fighters do on a daily basis. Officials experience wearing personal protective clothing and self-contained breathing apparatus (SCBA). They actively participate in search and rescue operations, auto extrication, operations of a ladder truck from its aerial platform, and extinguishing live fires inside the academy's burn building. The Congressional Fire Service Institute conducts similar training for members of Congress and their staffs.

Citizen Fire Academies

Citizen fire academies operate according to a very simple premise. They are designed to make citizens aware of the services their local fire department provides. They are also meant to increase fire and life safety awareness and promote a positive image of the fire department to the community. These programs are designed for adults to learn more about how the fire department is organized and operates. They are typically open to persons 18 and older who either live or work in the community hosting the program.

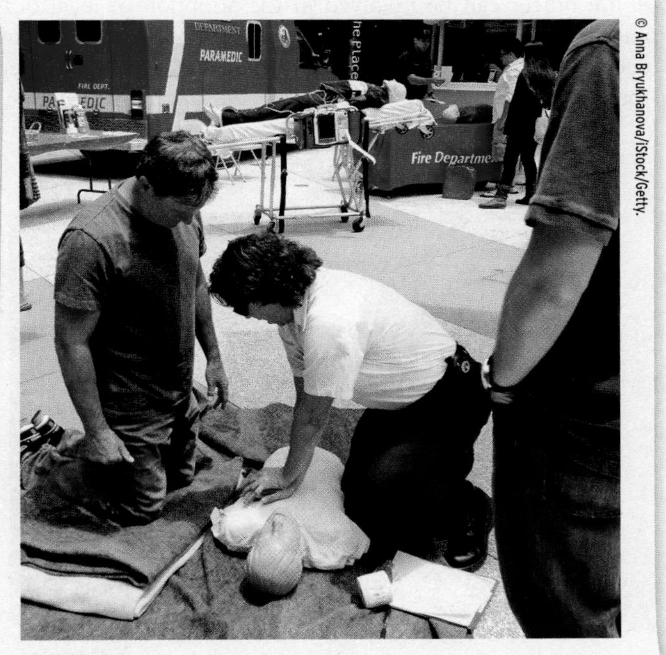

© Anna Bryukhanova/iStock/Getty.

FIGURE 6-2 One way a fire department can build a strong bridge into its community is to offer a series of CPR programs to its citizens.

Senior Programs

With the baby boomers now reaching their golden years, communities have developed many social programs that help provide an array of services. Many local or county health departments have senior programs on health and wellness that help sustain independent living in the community. Examples of these programs include home fire safety programs, fall prevention programs, senior health care, physical fitness routines, and Meals on Wheels for those who have limited abilities. Senior programs may be provided by the Council on Aging and/or senior support agencies. AARP is one national organization that promotes programs covering topics from health care to insurance to investment strategies. Triad, a team consisting of fire, law enforcement, and healthcare professionals, is another organization that assists seniors with preventive programs, neighborhood watches, and healthcare assistance information.

Fire departments have great opportunities to improve relationships with the senior demographic. Keep awareness of senior programs and other community programs high. Senior support can help departments at budget time, and, with the growing senior population, the tone of a community can change.

> **Fire Marks**
>
> **Citizen Fire Academy**
>
> One example of a citizen fire academy is found in Plano, Texas. The 10-week class meets one night per week at the fire administration building. Each night, citizen fire academy participants learn about a different aspect of the fire department, ranging from EMS to hiring and training fire fighters to special rescue operations. During the 10 weeks of the class, citizen fire academy participants also spend time at fire stations and riding on the engines to emergency calls. This is not a certification course.

First Aid

While many local healthcare providers such as hospitals and the local American Red Cross provide first aid programs, often the basic programs are conducted by the local EMS and fire department. Proactive EMS and fire departments provide community members with demonstrations of how to render first aid for people in need. The fact remains that the community has more of a natural trust in the emergency responders than in anyone else during emergencies.

The basic emergency care programs approved by the state department of health, which are suitable for delivery to the public, are delivered by the fire department. These may be Red Cross programs, state-mandated programs, or a version of nationally approved programs. First aid programs can also be conducted as a cooperative effort with a local hospital or medical center, the byproduct of which is an improved relationship with the facility-approved programs.

Emergency Management Team

Communities have developed emergency management teams to address the planning and needs of a community during any type of disaster. Since the inception of the 911 system, many communities have redesigned their emergency operation plans to have more of an all-hazard approach. Some of the key elements needed from the emergency management team are planning, mitigation, response, and recovery.

The chief fire officer must consider the role that he or she and his or her agency will be called upon to play in the local emergency management response structure. The most important factor in determining the long-term success of an emergency response to a major emergency is the performance of the local emergency management team. The team consists of individuals with a wide range of professions and trades. At a minimum, one should expect to have members from the following disciplines and groups on the local team:

- Fire
- Police
- EMS
- Emergency management
- Public works
- Engineering
- Local and regional communications
- Private sector groups such as the Red Cross and Salvation Army
- CERT
- Elected officials
- Public transportation

This team addresses all levels including federal response agencies such as the Federal Aviation Administration, the National Transportation Safety Board, the Federal Bureau of Investigation, the Secret Service, and others.

It is important to remember that the performance of an emergency management team on the day its services are really needed is the true measurement of its success, rather than any procedures or plans created or resources identified prior to the incident. This team either recovers from the emergency situation or fails in its efforts through its conduct during those first few precarious minutes and hours.

Today, many communities have received extensive training in an effort to provide a more comprehensive emergency plan. Hurricane Katrina brought a new awareness to communities about requesting help from local, state, and federal agencies and managing resources efficiently. (See the "Emergency Management and Response Planning" chapter.)

Public Health Department Team Member

As a society, we now have many members who work closely with the local public health department to address public health concerns. For example, after the first reported case of Ebola in West Africa in March 2014, public health officials and fire service leaders worked together on preparations for the response to the potential spread of the virus in the United States. The collaboration between the EMS and fire service and the rest of the public health team (e.g., providing education) brought a message of cohesion to community members and customers and demonstrated the diversity of today's fire service operations.

Law Enforcement

One of the most important intergovernmental relationships for the chief officer is with all levels of law enforcement **FIGURE 6-3**. Today, chief officers must be able to work within a unified command organization, often with several federal law enforcement agencies, state agencies, and county governments. Reaching out to the law enforcement community for support is especially important in those areas where law enforcement excels, including the following:

- Personnel and asset security
- Traffic management
- Criminal investigation
- Crowd control
- Communications support
- Investigation
- Intelligence gathering

Sustainability and Local Governments

Fire department sustainability occurs on multiple levels. A chief officer must constantly reassess a department's services and how to streamline and maintain them. Economic sustainability is often at the forefront of a department's (and the community's) concerns. Key activities in sustaining a department include reducing redundant and/or outdated services and ensuring the long-term security of revenue sources. Financial issues are discussed further in the "Budget and Finance Issues" chapter.

Code Understanding and Adoption

A fire officer can be involved in the adoption of local ordinances, local resolutions, state laws, fire codes, and building codes. Although many code models are nationally accepted, the need to research and redefine specific amendments to local

FIGURE 6-3 One of the most important intergovernmental relationships for the chief officer is with law enforcement.

laws will require interagency action with the local law enforcement, the special counsel, and even the county prosecutor. The state fire marshal's office can provide comprehensive insight into the state fire code and how it applies to the local fire executive. In many cases, the local chief fire officer must understand the statutes that provide the governance for the office he or she holds. Often, the chief officer can be held accountable for not enforcing the state and local law in a dangerous situation. It is therefore important that he or she seeks input from the department's attorney.

The chief officer must understand the ethics and compliance requirements in the enforcement of citations and must deal with political, social, and economic issues that can have far-reaching effects on the local community and the outcome of future development within the community. Balance becomes the key in deciding on important issues. (See the "Managing the Code Enforcement Process" chapter.)

Response Resource Management

Today's chief officer understands the value of having sufficient resources to effectively and efficiently mitigate the events to which he or she will be asked to respond. For the majority of incidents, most fire departments can safely handle response needs with internal resource allocation. For those incidents that exceed a department's internal resources, intergovernmental response agreements must be utilized to ensure that the resource pool is expanded, allowing incident commanders the ability to request the external assistance needed.

There was a time when fire departments were expected to maintain enough resources to handle their own incidents outside of a response that was considered a major event. With resource reductions brought about by financial constraints and government downsizing, many departments now rely on some level of response aid utilizing an external resource response. External responses are generally provided through some type of formal aid agreement authorized at the elected official level. This type of intergovernmental cooperation can be found in the form of both mutual and automatic aid. By pre-establishing these types of aid agreements, chief officers can provide their incident commanders with the resources necessary to mitigate their incidents safely. For many departments, mutual and automatic aid responses have gone from a luxury to a necessity, with many aid requests being made on a daily basis. Rather than concentrating on one type of aid, chief officers should look to include both mutual and automatic aid in their response planning. Flexibility is certainly the key when it comes to giving today's incident commanders the tools they need.

Mutual Aid

Sharing of resources in times of need has been a longtime staple for the fire service. Early mutual aid was provided through simple agreements that were negotiated with immediately surrounding departments. Mutual aid was typically reserved for use in the event that a department's own resources were depleted. Under mutual aid, incident commanders, upon determining the need for external assistance, would contact their dispatch agency and request the assistance of one or more of their mutual aid partners. The term *mutual* means that the agreement should work to the advantage of both parties: If I can send you an engine when you need it, then you should be able to do the same for me in time of need. The only exception may be the response of specialized apparatus such as an aerial device or ARFF unit for foam. Mutual aid served most of the fire service well for many years because, under normal circumstances, aid was needed only during the infrequent major fire.

As the value of mutual aid grew, some areas expanded its use in both scope and geographic size. One such example is the Mutual Aid Box Alarm System (MABAS) of Illinois. MABAS took the aid concept to a much greater level as it coordinated not only the normal mutual aid request but also evolved into an organization designed to develop, maintain, coordinate, and deliver specialized resource responses. Today, a MABAS response can be requested for the dispatch of hazardous materials and technical rescue teams, mass casualty response, and incident command support, as well as other functions. The success of the MABAS model has resulted in its endorsement by major fire service organizations and its implementation in other areas of the country. Michigan, a neighbor to Illinois, has embarked on its own adoption of the MABAS system. A project of the Michigan Association of Fire Chiefs, fire service leaders are seeking to develop MABAS divisions across the state. Success in building the MABAS network will lead to a more immediate delivery of needed resources to major incidents and disasters. In addition to providing the ability to coordinate a statewide response, many local jurisdictions have adopted the MABAS system for use in their county response planning, thereby consolidating and replacing older mutual and automatic aid agreements.

Automatic Aid

As the country's struggling economy caused a decline in local community resources of all types, fire departments looked once more to an expanded form of mutual aid. Motivated by personnel reductions, incident commanders found themselves with limited tactical choices, such as severely limited on-duty staff or a lack of needed specialized apparatus. This issue was not solely a career department concern. Many part-time and volunteer departments also experienced a decreased availability of fire fighters as recruitment and retention issues forced departments to operate shorthanded.

In rethinking the mutual aid concept, a simple question was posed: If asking a neighboring department for additional resources during times of need was a worthy request when on-scene resources were overwhelmed by the needs of the incident, then why not ask for the response of additional aid immediately upon dispatch, when it is already known that the department is understaffed? This immediate response plan was identified as automatic aid, and it soon became a popular method of ensuring that a rapid response force, capable of initiating on-scene operations safely, was en route at the time of the original call. Automatic aid agreements soon became a popular tool for departments that were looking to decrease the lead time required to get sufficient resources dispatched to the scene.

share information and thereby act in a close-knit, interactive manner. This concept grew in importance after it became one of the lessons learned from the attacks on September 11, 2001.

To have a solid communications program, the agencies involved need to meet on a regular basis to ensure that their system is meeting the identified needs of the constituent parties. A communications component needs to become a part of the community planning process, ensuring that the ability to interact grows with the increase in size and scope of the community and its service delivery demands.

In many communities, the fire department becomes the common link to emergency communications, and the chief officer manages and provides input into how the call for service is received. Emergency communication comes in many forms, such as local interagency cooperation with emergency management agencies, reverse 911 call centers, public information officers, 911 call answering centers, local churches, nongovernment organizations like amateur radio groups, or civic associations like the Red Cross or Salvation Army.

Political Communications and Information Sharing

Today, social media results in news traveling faster than anyone could have imagined only decades ago. Public officials are looking for information as soon as possible, but electronic media may lead to information being disseminated before proper notifications can be made. In many cases, the mayor, president of the local town commission, or county commissioner has to assure the public of their safety during an event or crisis. Public officials may introduce department heads who have a legal, jurisdictional, or operational responsibility in the incident or crisis at that time. Public officials must also introduce nongovernmental agency support along with other intergovernmental resources that will be involved in helping out during the time of the event or crisis. Direct reporting to the public by these officials can help provide current and, more importantly, accurate information to the public, thus helping to counter any inaccurate reporting through social media outlets.

Design Credits: Flames: © Photos.com; Chapter opener: © Glen E. Ellman; You Are the Chief Officer: © Jones & Bartlett Learning. Photographed by Kimberly Potvin; Voices of Experience: © Glen E. Ellman; Smoke: © Greg Henry/Shutterstock, Inc.; Chief Officer in Action: Courtesy of Scott Dornan, ConocoPhillips Alaska.

You Are the Chief Officer Summary

1. What is the basic difference between implementation of a mutual aid agreement versus an automatic aid agreement?

Under mutual aid agreements, resources are shared by request of the on-scene incident commander. Usually this request is made after arrival of first-due companies, the establishment of command, a completed size-up, and realization that additional help is needed. Occasionally the aid request is placed based on dispatch information received, visual clues seen during response, or known history of the location and typical incidents.

Under automatic aid agreements, departments predetermine the type of aid needed for various call types. This aid request is then entered into the community's dispatch system, with the requested aid resources sent at the same time as the resources from the department of jurisdiction. Automatic aid puts resources en route immediately without the need for a formal request at the time of the call.

2. How would the implementation of an aid agreement affect other areas of the department operation?

Aid agreements must take into account the fact that operational personnel from multiple agencies will need to work closely together on the same team. This close working relationship will necessitate that affected chief officers jointly plan for these responses by reviewing and updating policies and procedures and creating a common operational plan. Items to be reviewed include:

- Mayday responses
- Emergency fire fighter evacuation signaling
- Emergency communications
- Incident command
- Fire fighter rehabilitation

3. What are the lasting benefits of fire department involvement with an intergovernmental activity?

Intergovernmental activities allow the chief fire officer to develop lasting relationships with other organizations within local, regional, state, and federal agencies that have not traditionally interacted with the fire service. This allows the chief fire officer to learn about their operations, capabilities, special procedures, and resources. With this understanding, the chief officer can develop unique and creative approaches to community issues requiring an interagency response.

4. How does the fire department avoid creating a polarizing political issue when engaging with other agencies?

Focus on the situation to be addressed or the problem to be resolved. It is normal for agencies to embrace the "What's in it for me?" attitude when initially focusing on an intergovernmental effort. The chief fire officer can focus all of the parties on the objective and measurable outcomes of an intergovernmental effort and avoid a "Let's make it even" approach.

Wrap-Up

Chief Concepts

- Regardless of whether a fire department includes career, volunteer, or a combination of career and volunteer fire fighters, the chief officers charged with running the agency must remember that the citizens within the community are the true focus of fire department operations.
- Chief fire officers must be prepared to interact with a wide range of local, county, state, and federal agencies and must identify the key players at the local, state, and federal levels and within specialty groups with which the chief officer must interact.
- A properly structured fire department is part of a community emergency management plan that has been designed to allow for incidents to be handled at the local level.
- The key to influencing government lies in understanding how your local, county, and state governments work.
- Developing programs that improve and expand service and build partnerships with the public requires an intergovernmental approach.
- Policies are generally developed and adopted by the administrative sections of an organization, in consultation with the operational forces that will implement and utilize the procedures or protocols, and the procedures or protocols are developed and adopted for them by the senior staff and executive officers of a fire agency.
- External communications include emergency communications, political communications, and information sharing.

Hot Terms

Automatic aid The establishment through an agreement of the automatic dispatch of predetermined response resources from one community to another upon the dispatch of the initial call.

Emergency Management Assistance Compact (EMAC) A mutual aid agreement administered by the National Emergency Management Association.

Emergency management team A group of individuals with a wide range of professions and trades that can assist at an emergency.

Interoperability The level to which all emergency and nonemergency response agencies can share information and thereby act in a close-knit, interactive manner.

Mutual Aid Box Alarm System (MABAS) An agreement in which local, state, regional, and national response is requested through a decree for a large-scale emergency.

Nongovernmental organization An organization that is neither part of a government structure nor a for-profit business. It may receive funding from a variety of sources, including government, grants, private donations, and foundations.

Walk the Talk

1. Consider the demographics of your community and the social services provided by municipal and nongovernmental organizations. Identify three gaps between the community needs and the services that the fire department could provide.
2. Review your department's local aid agreements and consider areas that need improvement or updating. Write a proposal to the county chief's association outlining improvements to your existing agreements.
3. Make a list of on-scene operational policies and procedures that would benefit from a joint review with an eye toward adoption of a county-wide approach to policy and procedure revisions.
4. Conduct a review of your current operation to identify and evaluate potential public–private partnerships that may be expanded or implemented to address your community public safety concerns.
5. Investigate the availability of an existing CERT team in your community. If one is available, document its activation process and capabilities. If one is not available, develop a proposal to begin the development of a county-wide CERT team.

CHIEF OFFICER *in action*

Recent revenue reductions have impacted your department, resulting in administrative staff reductions and a moratorium on replacing retiring operational employees. At a recent conference you attended, a speaker from the Department of Homeland Security outlined several initiatives that utilized volunteers to assist fire departments with administrative and operational issues. After consultation with your senior administrative official, you have been given the green light to proceed with investigation of how these initiatives could assist you in providing service to the community.

1. Trained CERT volunteers can assist the department with all of the following, except:
   - **A.** fire safety tasks.
   - **B.** light search and rescue.
   - **C.** defensive-only firefighting.
   - **D.** disaster medical operations.
2. Assistance with computer data entry and filing can be provided by members of:
   - **A.** DMAT.
   - **B.** Fire Corps.
   - **C.** MABAS.
   - **D.** Any of the above
3. What level of fire experience should volunteers for a CERT program have?
   - **A.** At least five years of public safety experience
   - **B.** Retirement from a public safety agency
   - **C.** Any work with a government agency
   - **D.** None. They will be trained upon acceptance to the program.
4. The importance of CERT is especially proven:
   - **A.** during the first hours of a disaster when the community's normal public response agencies are overwhelmed.
   - **B.** during the recovery phase of the incident.
   - **C.** when the community's own public safety resources are able to work side by side with CERT members.
   - **D.** when mutual aid resources are unavailable.
5. Fire Corps is managed and implemented through a partnership consisting of:
   - **A.** the USFS and FEMA.
   - **B.** the Peace Corps and International Association of Fire Chiefs.
   - **C.** the National Volunteer Fire Council and the International Association of Fire Chiefs.
   - **D.** the United States Fire Administration and the National Volunteer Fire Council.

CHAPTER 7

Budget and Finance Issues

Fire Officer III

Knowledge Objectives

After studying this chapter, you should be able to:

- Describe how to manage a department's financial resources NFPA 6.4 NFPA 6.4.1 NFPA 6.4.2 NFPA 6.4.4 NFPA 6.4.5. (p 166)
- Discuss the types of funding available from public and private sources NFPA 6.4.2. (pp 166–169)
- Discuss the roles and types of budgets NFPA 6.4 NFPA 6.4.1 NFPA 6.4.2 NFPA 6.4.4. (pp 170–172)
- Explain the budgeting process NFPA 6.4 NFPA 6.4.1 NFPA 6.4.2 NFPA 6.4.4 NFPA 6.4.5. (pp 173, 175–176)
- Explain purchasing procedures in light of budgeting considerations NFPA 6.4 NFPA 6.4.1 NFPA 6.4.2 NFPA 6.4.3. (pp 176–178)
- Identify fire department funding sources NFPA 6.4.2. (pp 178–179)

Skills Objectives

After studying this chapter, you should be able to:

- Obtain funding from public and private sources NFPA 6.4.2. (pp 166–169)
- Create a budget NFPA 6.4 NFPA 6.4.1 NFPA 6.4.2. (pp 170–173, 175–176)
- Make department purchases in light of budgeting considerations NFPA 6.4.2 NFPA 6.4.3. (pp 176–178)

Fire Officer III and IV

Knowledge Objectives

After studying this chapter, you should be able to:

- Explain capital improvement planning NFPA 6.4.3 NFPA 6.4.5 NFPA 7.4 NFPA 7.4.1 NFPA 7.4.4. (pp 179–183)

Skills Objectives

After studying this chapter, you should be able to:

- Create a capital improvement plan NFPA 6.4.3 NFPA 6.4.5 NFPA 7.4 NFPA 7.4.1 NFPA 7.4.4. (pp 179–183)

MAYOR
KEVIN JOHNSON

This is your first budget as the new chief officer. The budget package has arrived on your desk and you are about to embark on an important first-time budget proposal that will set the tone for future years' budgets.

Preparing the deartment's budget is considered a high-priority task by your administrative official. The proper development of a budget is something that will greatly reflect upon your abilities as a chief officer. Although economic times are currently good, you know that your administrative official always takes budget preparation seriously and considers everything on the table for review in both good times and bad.

1. What is the best way to show that you are being proactive when it comes to capital expenditures?
2. What can be suggested within the budget proposal that can help offset or control the cost of running a fire department?
3. How can you better justify a budget request when it comes to general funds for items such as apparatus fuel, utility expenses, apparatus maintenance, and fire response?
4. How can an administrative fire officer improve competencies in budgeting and financial control?

Introduction

A fire department budget is the palette from which a department's existence, programs, and future initiatives are presented. In the past, some fire departments accepted whatever funds were granted to them from their municipality. They would make the allocated funds last as long as possible, shifting them among budget categories, until the next budget cycle began. If funding cuts were on the horizon, departments would simply reduce staffing to the new budget limitations. Lines between capital and operational funding were sometimes blurred. In tight budget years, chief officers would see funding taken away from capital purchases, delaying acquisition of equipment and apparatus in order to fund operational or general fund accounts. Additionally, when operating within a general fund budget, the chief officer is more or less in competition with directors of other general fund departments. This sometimes results in the transfer of funds among departments or functions within the governmental unit. Operating within a separate fund budget—e.g., fire authority—might eliminate the competition aspect of a general fund arrangement, but the fire authority budget has other issues to consider, such as the inability to acquire funds from other departments.

Fund-raising and recent grant opportunities have helped departments make ends meet, but these resources are becoming more scarce. Arguably, the most important parts of any fire department are its budget and the budget process used to achieve its mission. It does not matter whether it is a state, regional, or local municipality or a fire district, authority, or a private fire department; all need a clear, sound, sufficient, and stable budget from which to operate.

Being the administrator for a modern fire department is an awe-inspiring task. The decisions that affect resource allocation are growing more complex with each passing day. However, as the fire service moves through the early years of the 21st century, the charge to provide effective fire protection in an environment of diminishing financial resources has become extremely challenging. To understand the need for changes in the way fire and emergency service resources are used, it is important first to understand ways in which funds are generated, as well as the systems that allow for that management function to occur.

For chief officers, public sector administrators, and their political leaders, the changes have been coming in rapid succession. Just look at the changes that have come about because of terrorism (such as on September 11, 2001), adverse weather incidents (tornadoes, tsunamis, and hurricanes), and the economic downturn. Fire service leaders across the spectrum have been confronting changes of epic proportions. As part of the local government system, the fire service is a product of many of the rules, regulations, laws, and similar constraints placed on it by local government. To cope with the demands of this new reality, the fire service must be able to operate effectively within the budgetary system.

In the past, many leadership positions in the fire service were filled politically. Although there has been a movement away from political appointments, this method (which often results in high turnover rates) is still used in some departments. This type of centralized, political leadership created problems that may still be found in the operations of modern fire departments. Prior to the 1960s, little thought was given to the manner in which budgets were drawn up and resources allocated. The efficiency of fire department financial operations was not subject to any high degree of scrutiny. Perhaps the best way to describe the type of budgeting popular during that period was noted by a public administration expert of that era, Charles Lindblom. His research noted that, ". . . most public administrators simply added a little bit to last year's budget and hoped for the best. Budgets increased in a slow and incremental manner" (Carter 1989, 5).

This chapter discusses many of the important aspects of financial resource management that a chief officer must understand. It is critical for chief officers to become familiar with budgetary functions and to create budgets that provide an effective and efficient service for their communities. They must also be equipped to explain their budget requests to local government officials, the members of their fire department, and the public.

Most fire department resources are funded with public money, which is more closely scrutinized today than ever before. In some case, revenues may be generated from fund-raising activities such as donation solicitations. A creative chief officer, however, will find alternative ways of supplementing funds he or she receives from the governing body when necessary. TABLE 7-1 lists some examples of how the base budget may be supplemented by forward-thinking chief officers.

A chief officer must be concerned with justifying and managing the department's financial resources effectively and efficiently. A basic principle in municipal fire protection master planning is that the level of fire protection provided to a community is equal to the amount its citizens are willing to pay. Managers of progressive fire departments, however, inform their elected representatives of the probable costs of fire risk as it relates to fire protection at different levels of fire department funding. In other words, the fire department managers, as experts, provide estimates of the potential consequences that a community faces at the requested budget and at other budgetary levels. These consequences include alterations in service delivery, department activities, and level of community and department safety, among others.

Table 7-1 Potential Community Sources of Additional Funds

- Financial donations
- Equipment donations
- Fund-raising events
- In-kind donations by professionals
- Assistance from citizens with special skills
- Cost recovery
- Fines, permits, and fees
- Exchange of services with other public and/or private sector agencies

Source: Carter, Harry R. *Managing Fire Service Finances*. Ashland, MA: International Society of Fire Service Instructors, 1989. p 5.

Fire Officer III

Managing Financial Resources

The guideline for managing financial resources is to determine what needs to be done to ensure that the department will have the necessary funds and will administer them effectively to pay fair and equitable compensation to department staff members, maintain facilities and apparatus in full operating condition, obtain the needed equipment and supplies, procure additional facilities when needed, purchase additional apparatus, and refurbish existing apparatus to stay abreast of the needs of the community to be protected.

Specific issues for financial resource management include the following:

- Funds for the operating budget
- Funds for the purchase of needed facilities, apparatus, and equipment
- Responsible use of funds
- Elimination of wasteful practices in all aspects of operations
- Procurement policy

Origin of Public and Private Money

Today's fire service has taken on a new identity beyond that of providing fire suppression and emergency medical services (EMS) to its communities. It has become a business, and while this may be a startling revelation for some members of the fire service, the fact remains that we have long been regarded as a business by the members of our governing bodies and the citizens who pay the tab for our services. Furthermore, to understand why the fire service has received a greater amount of scrutiny, imagine that you are responsible for funding a business that is guaranteed to lose money. Regardless of which governmental level provides the agency's funding, the fire service administrator (chief officer) must at some point present a budget and justify funding for the department's operations. Consequently, ignorance of how government is financed can no longer be tolerated. If chief officers are not familiar with the financial operations of their government and the potential obstacles in increasing operational funding, they risk losing the annual budgetary battle. Such ignorance may also damage the fire department's reputation with the community and its governing body.

Today's chief officer must understand the source of his or her funding. If money is the fuel that fires the engine of government, then the manner in which the money is refined from the crude resources of the local jurisdiction should be examined. There are a number of ways in which government can gather the funds to operate, including:

- Taxation
- Assessment of user fees
- Impact fees
- Funds from other levels of government (grants, revenue sharing)
- Borrowing money from a variety of sources
- Funding from reserves accumulated from past budget surpluses

Another option is to seek funds from alternative sources outside a municipal government. As we have seen in each of the economic downturns over the past several decades, creative thinking and efforts have generated revenue-enhancing opportunities and/or reduced expenditures. Examples have been demonstrated in the creation of various public–private partnerships, cost sharing, mergers, and consolidation of services and efficiencies.

These methods, each with its own merits, may be used to increase the community's revenue based on the specific needs of the jurisdiction. There are, however, limitations and drawbacks inherent in each of these fund-raising mechanisms. All elements must be weighed when determining how to procure the necessary funds.

Taxation

Taxation is accepted by a community only if it is seen as equitable to the services those taxes fund. There is a wide range of taxes that could come into play in the development of budgetary funding. Depending on the state or region, there may be any or all of the following:

- Real estate and personal property taxes
- Sales taxes
- Income taxes

Each of these taxes impacts the individual taxpayer in a different way. The impact of real estate and property taxes varies as the value of property fluctuates. Most municipalities use several strategies to calculate their property taxes. The *millage rate* is the tax per dollar of assessed value of property (1 mill is $0.001). For example, a millage rate of 5 mills would generate $500 in property taxes for a home with an assessed value of $100,000. Assessed values are normally 50 percent of the real cash value of the property. Millage rates are data used by various community groups. For example, school boards may use millage rates within the school district to calculate local school taxes to be collected. When revenues fail to meet expenses, a government agency may put a millage request on the ballot for citizen consideration. In these cases, the voters are asked to approve an increase in the millage rate to fund government operations. Millage requests may be general in nature where, upon approval, the increased funding is placed into the general fund for distribution per the elected body's wishes. Another way is to ask citizens to approve a dedicated millage that can be used only to fund a specific operation (e.g., fire service).

The level of income tax payments varies with the income level of the individual being charged (as well as any deductions, credits, and exemptions). In many communities, local governments have approved the levy of local incomes taxes on citizens who live or work in the community. It is not uncommon to see the millage for residents set at one rate and the rate for nonresidents who work within the community half of that.

Everyone has to pay the sales taxes in cities, counties, and states that have created them. Sales taxes can be assessed on all purchases and in some cases on services as well. Some states and communities have exempted certain items such as food or clothing from sales taxes in order not to impact the lowest income residents of the community when purchasing basic needs items.

There are four generally accepted criteria for depicting the manner in which a taxation system is designed: yield, equity, neutrality, and ease of administration.

Yield

Yield is a critical element of taxation. A system of taxation is worthless if it fails to raise enough money to run the government. The assurance that enough money from taxation comes into the jurisdiction is the primary factor in designing a taxation system. An insufficient fiscal income stream may lead to borrowing to cover expenses. Most government officials have discovered that the creation of a broad-based system of taxation is the most effective means of funding government operations.

The real estate or property tax is one example of this type of system. Under this method of taxation, everyone who owns property pays. The homeowner, the business person, the industrial firm, and the mercantile operator all pay a tax based on the value of the property they own. This allows for a wider range of participation in the taxation system. In this way, taxes should be perceived as more equitable because everyone is sharing in the financial burden and everyone's share of the overall cost is proportional to the level of his or her holdings.

In addition to having as broad a base as possible, a system of taxation should have adequate income elasticity. It is crucial for the tax revenues to grow in a way that reflects the growth of income and growth in the need for services. Inflation is another factor affecting taxation, particularly in the world of income taxes. A good example of this could come from a community where there is a continuing growth in housing development, as well as increases in the population. Under such a positive growth posture, it might be possible to keep an even tax rate while increasing the revenue collected due to growth. Unfortunately, the opposite of this is also true. In communities where there is an outward migration of population, there may be a lowering of the tax base with fewer people to share the fiscal burden.

It is difficult to arrive at a proper level of elasticity. Municipal officials must monitor the growth/contraction patterns in their community closely. This can be done by reviewing the actions taken by local planning and zoning boards and building permit activity. As the level of activity increases, a period of growth can be forecast. As the level of activity tapers off, a period of financial contraction may be forecast. The chief officer should develop emergency risk planning data that can be compared to the growth patterns. In this way, progress may occur during periods of economic plenty. Of course, the reverse is equally true. Adjustments will need to be made as the actual fiscal realities play themselves out.

Equity and Neutrality

There will be fewer problems if each person feels he or she is paying a fair share of the cost of government services. Equity is a critical element in the success of any system of public sector fund-raising. No one likes to feel that he or she is being asked to shoulder an unfair share of any cost of government operations. When any level of government reaches the point at which government officials feel they can no longer raise the level of property and real estate taxation, the other types of fund-raising come into play. User fees are often chosen because they have the appearance of equity and neutrality, considering that each user pays the same fee. However, some residents may complain that a user fee actually equates to being taxed twice for the same service: once through the property tax system and a second time through the user fee. In communities where local income taxes are also assessed, user fees can come under additional criticism.

Ease of Administration

In discussing the general principles of taxation, one must not lose sight of the fact that taxes have to be administered. This imposes certain limitations on the fiscal process. The process must have clarity, continuity, cost-effectiveness, and convenience.

The primary goal of a national tax system is to generate revenues to pay the expenditures of government at all levels. Because public expenditures tend to grow at least as fast as the gross national product, taxes, as the main vehicle of government finance, should produce revenues that grow correspondingly. The facility of acquiring these revenues is known as ease of administration. Income, sales, and value-added taxes generally meet this criterion; property taxes and taxes on nonessential articles of mass consumption (such as tobacco products and alcoholic beverages) do not.

Assessment of User Fees

User fees are often used to place the cost of specific services with the individual or business that is requesting or utilizing the service. Examples of fire department user fees include:

- Fire code permit fees
  - Hot work permit
  - Hazardous materials use or storage permit
  - Special event public assembly permit
  - Fireworks display permit
- Suppression system inspection fee
  - Alarm testing
  - System bag test
- Fire watch standby
- Fire system plan reviews
- Site plan reviews

One reason local governments elect to use a user-fee scenario, such as the imposition of municipal transit fares or code enforcement permit fees, is the perception that each user of the service pays the same amount. A $2.00 bus fare is the same for all riders. However, is such a perception reality? A $2.00 user fee means one thing to a person earning $20,000 and another to a person earning $60,000. When put into perspective, it is easy to see how funding issues can generate considerable controversy. It is not uncommon for a community to experience conflicts between different neighborhoods, where differences in income level exist or user fees are considered out of line with what other communities are charging for the same service. Another type of user fee, also known as a *cost recovery charge*, is discussed later in this chapter.

Impact Fees

In some cases, the fire department may have the ability to asses an impact fee to a development. The premise behind an impact fee is that an investor who seeks to develop a project within a community needs to consider the financial impact that the development will have on governmental services provided to the development. For example, a large residential development may change the rural nature of the community and greatly impact the number of fire department responses. The community may ask the developer to assist with the funding necessary to construct, staff, maintain, or donate land for a new fire station that will be required because of the development. In another example, a developer may be required to fund the purchase of an aerial device because the proposed high-rise development is the first of its kind within the community.

Funds from Other Levels of Government

Many communities look to other levels of government for financial assistance. This may be because the community has projects and missions of such a substantial size that to fund them locally would place a tremendous strain on the existing system. In other cases, communities may need the funds for day-to-day operations. Historically, one of the additional sources of funding for local fire protection programs is the transfer funds available at the federal level, typically in the form of grants (discussed in more detail later in this chapter). Intergovernmental transfers of funds are also still prevalent in such areas as health and education.

Another source of funding for local services from another level of government is through the use of revenue sharing. Revenue sharing takes taxes collected by one level of government and reallocates those funds to the local level. The sharing of tax revenues can be done through statutory legislation or can be written into state constitutions.

Life Safety Initiatives

10. Grant programs should support the implementation of safe practices and procedures and/or mandate safe practices as an eligibility requirement.
16. Safety must be a primary consideration in the design of apparatus and equipment.

Borrowing Money

Borrowing to provide financial resources is another way to fund government operations. This is a common way to buy fire apparatus and equipment or to build new fire stations. The same holds true for all arms of government. Borrowing should not be for the purpose of providing day-to-day services, however. In the 1970s, New York City faced some serious financial issues because it had fallen into the habit of borrowing money to provide services rather than for acquiring equipment and apparatus. This practice led to severe problems; it was only through stringent financial controls, personnel layoffs, and program reductions that the city avoided bankruptcy. Given the volatile nature of the economy, such problems must always be on the minds of fire administrators and chief officers.

When considering borrowing, the cost of capital must be factored into the equation. This is a major concern in the choice of programs to be funded at the local level. If the funds are available from the budget or in the local capital reserve accounts, then the cost of capital is not an issue. However, when money must be borrowed for major projects, the costs associated with the projects become a critical element. Take, for example, the cost of borrowing money for a new fire

bid is awarded, the purchasing officer and approved vendor must agree on the procurement process (whether constructing a new station, building new apparatus, etc.). If the purchase is apparatus or equipment, training on correct usage will also be necessary and should be part of the process.

It is also a good idea to include contingency funds within the bid for unexpected expenses tied to the project. For example, when bidding a fire station, a contingency fund may cover expenses related to a change or upgrade in floor coverings. While contingency funds are common in facility construction projects, they are less common in the acquisition of apparatus. It would certainly be a prudent move if the chief officer included an amount of money in the apparatus specification as a contingency for unexpected changes to the apparatus specifications during the build process. Many chief officers have had bids for apparatus approved by their governing boards only to have to go back to the board for additional funding so that shelves, lights, and/or equipment that was overlooked can be added.

Chief Officer Tip

Maintaining Bid Integrity

The very nature of the bid process within government is to provide for fairness in the expenditure of tax dollars while ensuring the best price and compliance with minimum specifications as set forth by the department. It is important that the chief officer maintain the confidentiality and fairness of the bid process. The chief officer must avoid any behavior that might lead someone to believe that the bid process was flawed or misrepresented. The acceptance of gifts and/or trips from a vendor may be seen as trying to influence the chief officer when making bidding decisions, such as the development of specifications or the evaluation of submitted bids. The advanced release of projected bid cost received from one vendor and then provided to another prior to the official bid may signal favoritism in the awarding of a bid, which could be perceived as an unfair advantage.

You Are the Chief Officer Summary

1. **What is the best way to show that you are being proactive when it comes to capital expenditures?**

   Planning is the key to developing a good capital proposal. The development of a five-year capital plan can demonstrate to elected officials that you are being proactive and thinking about the future. There are many high-value equipment needs that support the department operations besides rolling stock and firefighting equipment. Planning for facility maintenance, appliances, copiers, computers, and the like is just as important as planning for apparatus replacement. That said, surprising your officials with a sudden request for the purchase of a $500,000 piece of apparatus is never a good idea. Instead, it may be better to identify all of your fleet issues, age, maintenance concerns, and usage and develop an apparatus replacement plan built over a long time frame, for example, 20 years. Your plan should include suggested replacement dates for each unit and the amount that will need to be budgeted within the capital program each year to save for the upcoming acquisitions. While the governing board or administrative official might modify or even reject the plan, you have shown your due diligence regarding the current and future needs of the department.

2. **What can be suggested within the budget proposal that can help offset or control the cost of running a fire department?**

   A good budget proposal should look at both sides of the equation: expenditures *and* revenue. Consider ideas such as cost recovery, grant programs, public–private partnerships, and other fees for service as revenue enhancements. While elected officials may not be quick to enact new fees in good economic times, being prepared and showing that you are examining the revenue side demonstrates good knowledge of the budget process. Researching and utilizing grant opportunities can also help offset the need for additional revenue. Applying for and receiving grant funds for new turnout gear can save capital funds for other projects. An overall review of operations can also produce savings by controlling expenditures, such as controlling thermostats, reducing water usage, controlling lights and other items that consume electricity, and reducing fuel expenses by eliminating unnecessary trips with apparatus.

3. **How can you better justify a budget request when it comes to general funds for items such as apparatus fuel, utility expenses, apparatus maintenance, and fire response?**

   Everyone knows that some budget numbers are really educated guesses. We project fuel cost, maintenance expenses, utility usage, number of responses, and overtime pay. One way to help justify the numbers presented is to examine budget history for these types of line items. Calculating average expenditures for particular items over three- and five-year periods can help smooth out the low and high years. While not foolproof, it is certainly better than a guess.

4. **How can an administrative fire officer improve competencies in budgeting and financial control?**

   An administrative fire officer can improve competencies in budgeting and financial control by attending a college-level course in fiscal administration, municipal budgeting, or public sector economic analysis. For example, the National Fire Academy offers Fire Service Financial Management (R333) as a two-week on-campus course.

Wrap-Up

Chief Concepts

- The most important parts of any fire department are its budget and the budget process used to achieve its mission.
- The guideline for managing financial resources is to determine what needs to be done to ensure that the department will have the necessary funds to:
  - Pay fair and equitable compensation to department staff members
  - Maintain facilities and apparatus in full operating condition
  - Obtain the needed equipment and supplies
  - Procure additional facilities when needed
  - Purchase additional apparatus
  - Refurbish existing apparatus
- Funding can come from:
  - Taxation
  - Assessment of user fees
  - Impact fees
  - Funds from other levels of government
  - Borrowing money
  - Adding funds to surplus for future purchases
- By laying out the needs of the fire department in a systematic manner, a fire department chief officer and administrator can show the public and the members of the governing body what the fire department will require to achieve its operational goals.
- Fire department officers often must justify the expense of an item. There are two aspects to such justification: showing need for the item and showing why the requested funds are the best way to satisfy the need.
- Budgeting is a four-part process that includes the following:
  - Formulation
  - Transmittal
  - Approval
  - Management
- Clear and established purchasing rules are the backbone to an easy and efficient program. Whatever systems and programs are put into place, the rules need to be clearly defined and follow all department policies and municipal, state, and federal laws.
- Securing alternative funding is a practical income management measure for fire departments to meet essential operational needs and expenses.
- Running a fire department is a capital-intensive operation. Capital funding requests can address many department needs, including the addition or replacement of facilities, facility repairs or remolding, and equipment—from protective gear to AEDs to extrication devices.

Hot Terms

Approval The crucial phase of the budgetary process wherein the governing body weighs the merits of a given request and balances the differing requirements of all agencies applying for funding. Those requests found to have adequate justification and operational merit are funded within the available resources of the governing body.

Bid laws Laws that require competitive bidding to ensure that the lowest possible price is received.

Budget An itemized summary of estimated or intended expenditures for a given period, along with financing proposals for funding them.

Budget process The process used by an agency to develop and present its annual financial requirements to its governing body for approval and use.

Budget spreadsheet A listing of how much money was budgeted in the previous year, how much was actually spent, and how much is requested for the current year.

Capital budget An itemized summary of those items where the cost must be spread over a number of years.

Ease of administration The facility with which funding is provided to the operational elements of a governmental entity.

Equity The element of perceived fairness.

Expense budget A plan for future operations, expressed in financial terms. Sometimes called the operating budget.

Expenses Charges incurred or expenditures of money to support various programs and their elements within the organization.

Formulation The budgetary phase during which the budget is created within the agency.

Government bonds Promissory notes issued by government to raise funds for operational necessities of a major, higher-cost nature.

Impact fee A fee charged to the developer of a project by the local governmental entity to help cover the cost of providing service (e.g., fire, police) to the new development once completed.

Income tax Revenue request based on the level of income an individual, corporation, or organization generates.

Line-item budget A budget system in which amounts are budgeted in separate accounts based on the department, division, category, and specific expense type. Each expenditure is then applied to a specific line of the budget when submitted for payment.

Millage request A method of funding based on increases in tax millage rates. It is approved through citizen vote.

Neutrality The need to remain neutral as to the manner in which funds are generated by government.

Program budget A type of financial planning document wherein managers look at their organization in terms of actions and activities.

Real estate and property taxes Financial demands to fund operations, made upon those who own property in a given jurisdiction.

Revenue sharing Sharing of collected funds between government agencies.

Sales taxes Revenue requests based on the sale of goods or services in a municipality, county, or state.

Taxation The levying of revenue requests on those charged with supporting government operations.

Transfer funds Fiscal resources received from another, usually superior, level of government.

Transmittal The stage in which the budget, as developed by the agency, is forwarded through the appropriate mechanism provided by the governing body.

User fee A fee charged to those using municipal services.

Yield The amount of money raised by the method of fund-raising used by government.

Walk the Talk

1. Develop a spreadsheet that shows both the three-year and five-year average expenditures for each of the line items in the department's budget.
2. Develop a five-year capital plan for major department expenditures relating to facility and equipment needs. Include things such as heating and cooling units, roofs, station furniture, kitchen appliances, and other items not normally recognized as fire related.
3. Develop a replacement proposal for a front-line fire engine that is 20 years old and beginning to experience increased maintenance costs. There is no current replacement plan for this purchase.
4. The loss of the largest employer in your community will reduce next year's municipal budget by 15 percent. Outline how your department would function with 85 percent of the current year's operating budget.
5. Describe three revenue-generating activities that your fire department could implement to offset loss in municipal revenue. Identify local, state, or federal administrative regulations that need to be considered.

CHIEF OFFICER *in action*

Your community is facing hard times resulting from several manufacturers leaving town. Jobs have been lost, homes have been foreclosed, and property values have declined. An economic recovery is projected, but not for at least five years. In the meantime, your budget process is just beginning and all department heads have been asked to review budgets and prepare for modification to deal with the new fiscal realities. As you sit down at your desk with a copy of the department budget, you consider your future options.

1. In hard times, what service would best be considered as nonessential and listed for possible elimination?
   - **A.** EMS non-transport service (You are trained to the first responder level and respond in conjunction with a hospital-based ALS transport service.)
   - **B.** Residential smoke detector installation (Detectors are supplied free under a grant program.)
   - **C.** Fire department first responder standby at high school sporting events (You utilize off-duty personnel and pay them to stand by with an engine for medical emergencies. The local ambulance is also there to assist.)
   - **D.** School station tours (Local elementary schools bring their students to tour the fire station and hear fire prevention messages from on-duty fire fighters.)
2. Which additional revenue source would require a vote of the citizens?
   - **A.** Enacting a cost recovery charge
   - **B.** Raising the property tax millage rate (Your millage cap is set by city charter at 10 mills and you are currently assessing 6.8 mills.)
   - **C.** Beginning a new dedicated fire services millage
   - **D.** All of the above
3. If there is no separation between capital revenue and general fund monies, which expenditure item would be best recommended for elimination to meet the current fiscal year budget requirements?
   - **A.** Delay purchase of new engine, replacing 20-year-old unit.
   - **B.** Lay off full-time staff.
   - **C.** Delay remodel of the current fire station to update office space.
   - **D.** None of the above; all are important projects.
4. As you begin the budget process, how much input from your staff should you utilize in preparing the proposal?
   - **A.** None, since it is best to complete the budget yourself when cuts are needed.
   - **B.** It would be best to delegate the budget proposal to your deputy chief and let him or her assess the operations that should be reduced or eliminated since that is his or her area of control.
   - **C.** It would be helpful to form a budget committee with representatives of all employee groups for input on the budget proposal.
   - **D.** It is better to conduct informal conversations with individual employees to gather intelligence on what they may be thinking regarding possible acceptable cuts.
5. Economists are predicting that your community will continue to have an eroding tax base with a three to five percent decline in annual revenues from property taxes, personal income taxes, and general sales/gross receipts tax. Your long-range plan should:
   - **A.** identify the programs, assets, or activities that will be eliminated.
   - **B.** identify and budget for expanded programs from a proposed innovative revenue-generating program.
   - **C.** anticipate continual but minimal growth, which translates to a two and a half percent revenue increase in each year.
   - **D.** maintain existing services. Anticipate that further efficiencies will be found within each budget year to cover any shortfall.

CHAPTER

8 Strategic Planning

Fire Officer III

Knowledge Objectives

After studying this chapter, you should be able to:

- Discuss the history of strategic planning in the fire service NFPA 6.1.1. (p 190)
- Discuss the importance of information management NFPA 6.1.1 NFPA 6.1.2 NFPA 6.4.6 NFPA 6.6.3. (pp 190–191)

Skills Objectives

There are no Fire Officer III-only skills objectives for this chapter.

Fire Officer III and IV

Knowledge Objectives

After studying this chapter, you should be able to:

- Discuss the concept of strategic planning NFPA 6.1.1 NFPA 6.1.2 NFPA 6.4.6 NFPA 6.6.3 NFPA 7.4.3. (pp 191–193)
- List and describe the steps in the strategic planning process NFPA 6.4.6 NFPA 6.6.3 NFPA 7.2 NFPA 7.4 NFPA 7.4.1 NFPA 7.4.3 NFPA 7.4.4. (pp 193, 196–199)

Skills Objectives

After studying this chapter, you should be able to:

- Create a strategic plan. (pp 191–199)

You Are the Chief Officer

It is your first senior staff budget meeting as the chief officer of the department. You have reviewed the fire department's past budget request and have noticed that previous chief officers have been unsuccessful in obtaining requested increases for additional staffing, updated equipment, and facility improvements. Staff has reminded you that the department has been trying to replace the reserve aerial ladder for a dozen years. It is obvious to you that the department needs a new approach for future budget requests.

A snapshot of department responses shows that service demands have risen in 8 of the past 10 years:

- 2005: 1256 responses
- 2006: 1349 responses
- 2007: 1310 responses
- 2008: 1789 responses
- 2009: 2013 responses
- 2010: 2259 responses
- 2011: 2352 responses
- 2012: 2319 responses
- 2013: 2482 responses
- 2014: 2654 responses

After thoroughly discussing options with your staff, it is decided that the best way to proceed is to develop a long-range strategic plan. The plan will be used to identify community risk, the adequacy of current levels of service, and future service trends. It is hoped that the plan will assist in convincing the elected officials to support future budget requests.

1. How does a managing or administrative fire officer match the vision with the available resources?
2. What statistics would be helpful in demonstrating current and future fire service needs?
3. What happens when the fire department strategic plan is rejected by the authority having jurisdiction?
4. The fire chief has been trying to replace the reserve aerial truck for more than a decade. When does a chronic unmet need become a normal operating condition?

Introduction

One of the critical elements fire department officers at the Fire Officer III and Fire Officer IV levels will face is the requirement to conduct a needs assessment for their fire departments. To provide a proper level of service to their district, fire departments should tailor their operation to the actual service delivery levels required by their community. Once the needs in their community have been identified, the next important task becomes the development of a strategic plan for future operations within the agency. These analyses and plans must reach out to other affected parties in the community. The community fire protection system comprises groups such as the public, government officials, business owners, and other city/town/township agencies. Any type of planning conducted in a closed environment is doomed to fail; it is important to meet with as many community groups (stakeholders) as possible in order to develop interoperable service agreements.

The purpose of a strategic plan is to provide a road map by which an organization can realize its vision of the future by following a set of goals or initiatives that accomplish the organization's mission. A good strategic planning process affords stakeholders the opportunity to become involved and participate, enhances teamwork and cooperation, and provides a basis for measuring performance.

Goals, measurable objectives, performance measurements, and strategic plans become instruments for managing and tracking resources and progress. A well-crafted strategic plan, guided by good management and executed by committed personnel, will ultimately translate to improved efficiency and better quality of services being delivered.

Fire Officer III

History of Strategic Planning

Although it may seem that chief officers have always recognized the need for planning when preparing department budgets, today's fire service leaders have now begun to place a greater emphasis on the importance of (and need for) strategic planning, involving topics such as staffing, station placement, and services provided. Interestingly, the idea for strategic planning is not as new as some would expect. In fact, its focus started with the Wingspread Conference in 1966, when the concept of planning for future operations began to grow in importance. The report of the conference proceedings specifically stated, "The traditional concept that fire protection is strictly a responsibility of local government must be reexamined" (Johnson Foundation 1966, 15). Prior to this point, no one spoke of fire protection in any way other than localized terms. This was the first look at the fire service as something that lived within a larger world.

In 1969, federal legislation was passed authorizing the creation of the National Commission on Fire Prevention and Control. It was to be located within the Department of Commerce, and its charge was to determine the extent of America's fire problem. It was then charged to develop recommendations for improving the nation's response to fire and its related issues of death, injury, and property damage. The report, titled *America Burning*, was issued in 1973. It assessed a number of areas, such as the following:

- The fire services
- Fire and the built environment
- Fire and the rural wildlands environment
- Fire prevention
- Programs for the future

One of the primary recommendations to come out of this report was the call for a fire service commitment to a concept called *master planning*. The report defined these master plans as documents that "should set goals and priorities for the fire services, designed to meet the changing needs of the community" (Bland 1973, 29). At this point, the fire service community began to consider that a need may exist to organize fire protection, as well as to develop a series of understandable guidelines.

Ten years after the first Wingspread conference, the Johnson Foundation in Wisconsin sponsored another conference. The report from that event broadened the intent of the original 1966 commentary. In this case, a number of new concepts were mentioned, including (Johnson Foundation 1976, 11–17):

- The state level of government may have to make a renewed commitment in dealing with the fire problem.
- The fire service should approach the concept of regionalization without bias.
- Fire departments should thoroughly analyze new demands being placed on them before accepting new responsibilities.

These words of future orientation were echoed and amplified by the Wingspread III Conference in 1986. In its report, government decision makers were urged to develop and implement better criteria in determining the best way to develop a cost-effective approach to local fire protection delivery systems. The report participants went on to note that the traditional role of fire departments was changing (Johnson Foundation 1986).

Yet another conference was held in Dothan, Alabama, in 1996. The International Association of Fire Chiefs (IAFC) hosted this long-range, high-level planning meeting. It was determined by the attendees that the pace of change in the fire service was increasing and that the old goals needed to be reexamined. The findings amplified the findings of the earlier documentation. However, a number of important new thoughts emerged from the session. To allow for the proper planning for fire protection delivery systems, it was determined that nationally recognized standards that would provide guidance for such critical issues as the following should be adopted (Wingspread IV, 2):

- Types of services to be provided
- Operations
- Deployment
- Evaluation criteria
- Response times

The fire service has traveled quite a road from its former reactive approach to the goal of becoming a proactive force—a force that is continually preparing for the future. The journey is not complete, but through proper strategic planning, fire officers can better prepare their fire service members and the greater community for the challenges that always lie just ahead.

Information Management

As much as increased responsibilities and decreased funding necessitate good planning, planning necessitates good information management. Because most everything a fire department does involves location, a geographic information system (GIS) is an effective way to manage the information a fire department uses to build a strategic plan. Many communities already have or have access to an enterprise-wide GIS that can provide critical information resources to aid the planning process. Regardless of whether your jurisdiction already has a GIS, the strategic planning process is a good way to begin building an information management platform that will allow you to take advantage of GIS benefits.

From an integrated executive-level perspective, an information management strategy that allows personnel across the department to share data among various systems greatly enhances the department's effectiveness and efficiency. Geography is the common characteristic on which data can be organized, allowing the different applications to access the same data about one specific point of interest, whether it is a person having a heart attack, a building on fire, or a staging area during a major disaster. GIS data can help graphically display response-type data, such as response times, response districts, mutual and automatic aid responses, and water supplies.

Fire Officer III and IV

Understanding Strategic Planning

The term strategy refers to a plan of action designed to achieve a particular goal. Fire officers use strategies when developing a firefighting action plan. The fire service has utilized various planning processes since the 1960s. From the 1960s through the 1980s, fire service planning consisted mostly of the development of incident management models, preincident planning, mutual aid agreements, and station location planning.

Fire service planning models evolved with establishment of the fire agency accreditation process by the Commission on Fire Accreditation International (now a division of the Center for Public Safety Excellence). To meet accreditation requirements, fire agencies are required to have a strategic plan.

Today, fire agencies commonly engage in the following three types of planning:

1. Tactical/operational planning
   - Preincident planning
   - Operational and shift planning
   - Demand-based resource planning
   - Local mutual aid planning
   - Regional mutual aid planning
2. Strategic planning
   - Focused on short-term achievements
   - Moves the organization toward the accomplishment of a common vision
3. Master/long-range planning
   - Long-range results that are performance oriented
   - Concentrates on facilities, apparatus, and staffing
   - Closely tied to cost projections and revenue planning
   - Linked to community planning

There are a number of ways to approach strategic planning. In the private sector, strategic plans are often created by top management and contain proprietary information. However, when public funds are utilized, it is important that the planning process be transparent and open to public scrutiny. Consequently, the strategic planning process should contain the following elements:

- Citizen involvement from the community(ies) served
- Member input from all levels of the organization
- Fire service input from nearby or adjoining organizations
- A strategic planning team representing a cross-section of the organization
- Other planning team members for consideration:
  - Members from other internal municipal agencies
  - Key community business leaders
- Buy-in, preferably early, by the organization's key leadership (e.g., elected officials, chief administrative officer, fire chief)
- Public presentation of the proposed and final versions of the plan
- Availability of the final plan to the public and all organization members

Chief Officer Tip

Transparency

When public funds are utilized, it is important that the planning process be transparent and open to public scrutiny.

The most successful strategic planning effort (process) involves both internal and external stakeholders and customers. The process provides for customer input and is driven by the central theme of what is in the best interest of the citizens served. The strategic planning process is designed to evaluate needs, focus on internal and external customers, and identify issues from the stakeholders up rather than the mission statement down.

The strategic plan should contain the following elements:

- A vision that stretches the organization and encourages excellence
- A mission that identifies the organization's core business functions
- Values that identify organizational and individual behaviors
- Goals and objectives that encourage and embrace involvement, participation, and teamwork
- Prioritization of the organization's needs
- Identification of performance measures for achieving desired and predictable outcomes

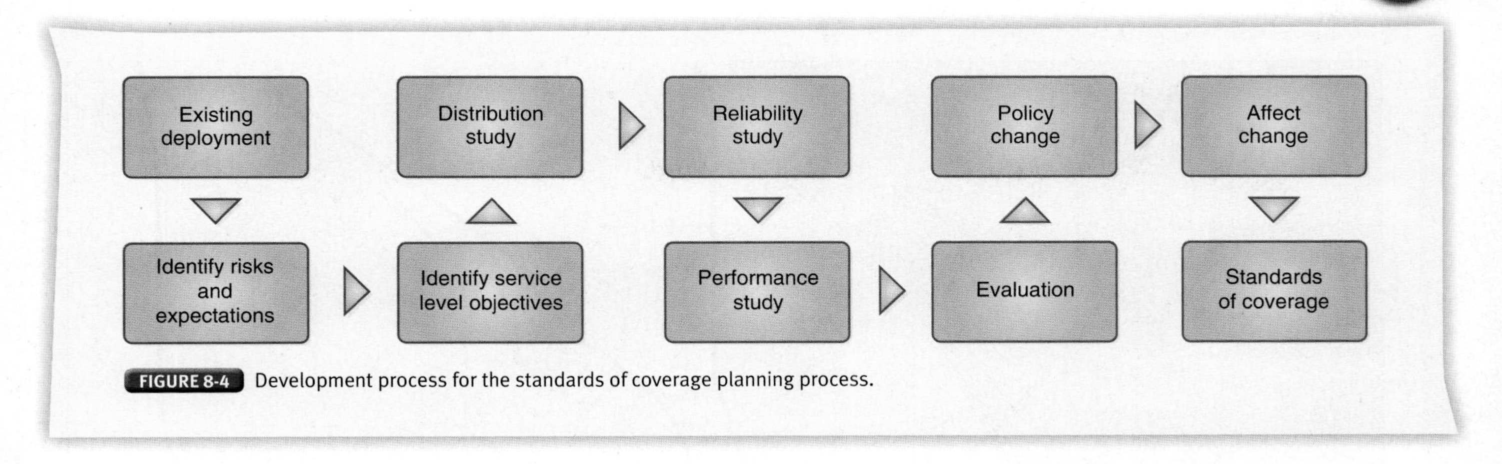

FIGURE 8-4 Development process for the standards of coverage planning process.

than in a low-risk building with a low occupancy load. More resources are required to control fires in large, heavily loaded structures (more to burn) than are needed for fires in small buildings with limited contents.

Most emergency medical incidents require the quick response of single fire crews to limit the suffering and to rapidly intervene in life-threatening emergencies. Small incipient fires need prompt response. . . . This is why distribution planning strives for equity and timely service.

The Plainfield plan utilized the format provided in *CFAI Standards of Cover.* Because the ultimate goal was to obtain support of the plan from elected officials, an attempt was made to provide a cost-versus-benefit analysis for providing emergency response capability within the community, allowing elected leaders to make an informed decision on the level of fire services provided to township residents.

In the development of this plan, the department looked to continue their focus on meeting a goal for total response time of 5 minutes (1 minute turnout time and 4 minutes travel time) 90 percent of the time. This allowed the developers to plot the needed potential new station locations to meet this benchmark. The response time goal was developed while taking into consideration the time to flashover on a fire call and brain death for a medical response. In the case of a fire response, flashover represents a significant increase in fire growth corresponding with an increase in needed fire flow and resources. Because it is known that the required response force for maximum risk areas is greater, a 9-minute response time, 90 percent of the time, was recommended as the target. It was established that an effective response force for maximum risk areas would be able to meet a needed fire flow of 900 GPM (57 L/sec) for firefighting. This response force level would also be able to handle a three-patient medical emergency. It was recognized that the effective response force would not necessarily reflect the total needed resources should the situation escalate or if the response were to a building preplanned for a worst case scenario of 4000 GPM (252 L/sec) fire flow.

The planning zones identified in the plan were established using the department's first-due engines FIGURE 8-5. These zones were further defined using the township's natural jurisdictional boundaries, including a highway (US 131) and the Rogue River. The department utilized the multi-county Regional GIS system (REGIS), which was driven by ArcView software, to create overlays including station territories, response grid map, fire hydrants, and National Fire Incident Reporting System (NFIRS) data (incident type).

Appendix C contains the data, mapping, and staffing recommendations from the standards of coverage report.

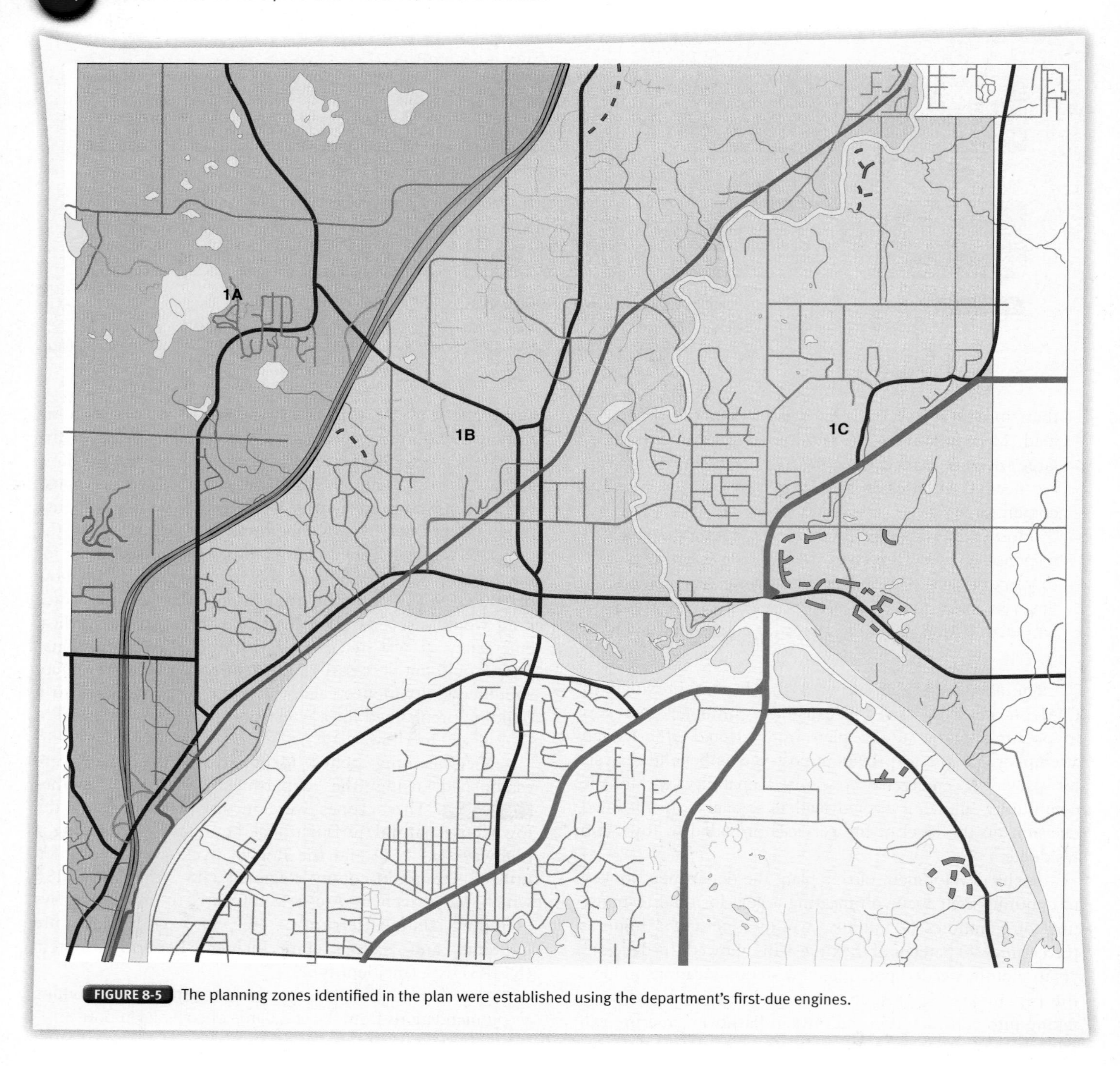

FIGURE 8-5 The planning zones identified in the plan were established using the department's first-due engines.

You Are the Chief Officer Summary

1. **How does a managing or administrative fire officer match the vision with the available resources?**

   A managing or administrative fire officer must focus on what is to be accomplished rather than on the specific tool to use. The department staffs two truck companies and has one spare rig that will probably fail its next required recertification test. Consider other ways to provide a fill-in aerial when Truck 1 or Tower 2 is out of service for maintenance or repair.

2. **What statistics would be helpful in demonstrating current and future fire service needs?**

   Having the data to back up projected service needs and recommendations is an important component of any strategic plan. The following items should be considered for review in a strategic plan:
   - Annual response statistics: total responses, fire responses, EMS responses, other responses, average response times
   - History of staffing levels: staff types, station staffing, response staff
   - Annual budget expenditures
   - Emergency response fleet data: apparatus type, age, acquisition and replacement cost, usage, expected replacement date
   - Facility data: location, age, utility expense, condition, needed upgrades, recommended upgrades

3. **What happens when the fire department strategic plan is rejected by the authority having jurisdiction?**

   If the department's strategic plan is rejected, the executive or administrative fire officer develops alternative plans that best meet the department's goals within the available resources. Clearly communicate the outcomes of alternative plans with respect to the original strategic plan—for example, the impact that the closing of a fire company has on response time and on-scene tasks.

4. **The fire chief has been trying to replace the reserve aerial truck for more than a decade. When does a chronic unmet need become a normal operating condition?**

   This is a political as well as pragmatic question. If the request has been rejected by two different municipal leadership teams, or in four annual budget requests, it is time to revise the strategic plan and to consider reserve Truck 12 disposed of without replacement.

Wrap-Up

Chief Concepts

- One of the critical elements that fire department officers at the Fire Officer III and Fire Officer IV levels face is the requirement to conduct needs assessments for their fire departments.
- The fire service has traveled quite a road from its former reactive approach to the goal of becoming a proactive force—a force that is continually preparing for the future.
- Fire agencies commonly engage in three types of planning:
  - Tactical/operational
  - Strategic
  - Master/long range
- One aspect of a fire department's various functions is to prepare for unwanted events. Another aspect involves using preventive efforts to mitigate the possibility of an incident occurring.
- As much as increased responsibilities and decreased funding necessitate good planning, planning necessitates good information management.
- There are five basic steps in creating a strategic plan:
  1. Preplan.
  2. Acquire planning data.
  3. Prepare for a strategic planning retreat.
  4. Conduct the strategic planning retreat.
  5. Prepare a written strategic plan.
- The specific elements of a strategic plan include:
  - Mission statement
  - Values
  - SWOT
  - Goals or strategic initiatives
  - Objectives
  - Organizational performance measures

Hot Terms

Interoperable The ability of one system (or organization) to interact with other systems (organizations). Generally used to describe the ability of communications systems from different organizations to achieve solid working relationships.

Objectives The basic part of the planning process. These are the tasks that must be accomplished to meet the goals of the organization and thereby achieve the mission for which the organization was formed.

Strategic planning The process by which an organization defines its course of action for the future.

Strategy The overall model of organizational reasons for existence as set forth in a variety of policies, programs, and procedures that define what an organization is, what it was created to do, and how it will accomplish its goals.

SWOT The strengths, weaknesses, opportunities, and threats that an organization will face both internally and externally.

Turnout time The time from having enough information to respond until the apparatus is placed into motion.

Walk the Talk

1. A recession resulted in draconian cuts in staffing and service for many communities. Speak with an executive or administrative chief officer in such a department to see what their strategic plan is for the next 5 years.
2. The nature of 9-1-1 calls has changed. There are more vehicle crashes, hazardous situations, and medical assist responses. How can you preserve an effective initial firefighting force that will need to handle less than 15 percent of the 9-1-1 responses in a year?
3. Strategic planning is an important tool for any size fire department but it also requires a commitment of time and resources to properly complete a strategic plan. Draft a proposal that will convince your supervisor that beginning the strategic planning process is a necessary and wise investment for your department. In your proposal, include what steps will be taken, who will be involved, how long it will take, and what you hope to accomplish.
4. NFPA 1710, *Standard for the Organization and Deployment of Fire Suppression Operations, Emergency Medical Operations, and Special Operations to the Public by Career Fire Departments* establishes a baseline for initial structural fire attack: Deploy 15 to 17 fire fighters to arrive within 8 minutes of travel time. In developing a local strategic plan, the labor union wants to mandate four-person fire crews. A first alarm brings three engines, one aerial, and a chief officer. As the fire chief, you would like the flexibility to assemble 17 fire fighters through a wider collection of staffed rigs, including a six-person aerial, three-person engines, and two-person quick attack units. Your first alarm would be two quick attack units, two engines, one aerial, and a chief officer. How would you approach this issue and what data could you use to back up your plan?
5. Develop a map of your response district that identifies locations of emergency responses, fire station locations, target hazards, and water supply availability.

CHIEF OFFICER *in action*

"The good news is that no one has life-threatening injuries." Assistant Chief Arrow was speaking with a group of rain-drenched command and administrative officers standing in a downtown street. While responding to a structure fire, Tower 2 collided with a delivery truck, struck the side of a downtown building, and overturned. "The bad news is that the tower is probably totaled."

Returning to their budget discussion, Assistant Chief Arrow noted that the Riverside Volunteer Fire Department, Inc., owns two aerials. The legal agreements that came from the 1967 annexation include a provision that the Riverside Volunteer Fire Department controls who operates their equipment and where they are quartered. The battalion chief is tasked to develop an aerial services coverage plan as part of an upcoming budget proposal to the city council.

1. Identifying the number of emergency incidents that require the unique capabilities of aerial ladders and platforms is an example of the __________ activity in strategic planning.
   - **A.** forming strategic view
   - **B.** setting objectives
   - **C.** crafting a strategy
   - **D.** evaluating impact
2. After looking at the annual report on truck company activity, the mayor's staff suggests that Tower 6 be moved from the River's Edge fire station to the Hillside fire station. Explaining the legal relationship between the Riverside Volunteer Fire Department, Inc., your fire department, and the 1967 annexation is an example of:
   - **A.** an external review of the community's needs.
   - **B.** an internal review of regulatory impact.
   - **C.** an external review of demographic needs.
   - **D.** an internal review of fire department performance.
3. The budget director suggests that the purchase of a quint vehicle could be supported by the city if it results in the combining of the crews from Engine 2 and Tower 2. The net reduction of 12 positions offsets the cost of a new aerial. This is an example of:
   - **A.** redefining a strategic vision.
   - **B.** completing a SWOT analysis.
   - **C.** a best and final offer.
   - **D.** crafting a strategic plan.
4. Looking at the annual report on truck company activity, the Citizens fOr $ensible Taxation (CO$T) suggest closing Truck 2. Because it averages just 2.1 responses per 24 hours, the cost of maintaining a dozen city employees on a million-dollar vehicle does not make sense. This input would show up when:
   - **A.** redefining a strategic vision.
   - **B.** completing a SWOT analysis.
   - **C.** completing a community risk hazard analysis.
   - **D.** crafting a strategic plan.
5. The president of the fire fighters' labor association provides statistics showing that the most rescues and removals of citizens from burning buildings are performed by Truck 2. Reviews of the structure fire reports from the past 5 years indicate that Tower 2 has rescued or removed 27 occupants from 465 structure fires, whereas Truck 1 has rescued or removed 9 occupants from 542 structure fires. This input would show up when:
   - **A.** redefining a strategic vision.
   - **B.** completing a SWOT analysis.
   - **C.** completing a risk hazard analysis.
   - **D.** crafting a strategic plan.
6. The Downtown Building Owners and Managers Association (DBOMA) wants assurances that Truck 1 will remain staffed and assigned to the headquarters fire station. It notes that two-thirds of the downtown buildings more than five stories tall do not have sprinkler systems or sophisticated fire control rooms, including the original section of Memorial Hospital. The chief wonders if the association will help pay for a new aerial truck at Station 1. This input would show up when:
   - **A.** creating a public/private partnership.
   - **B.** completing a SWOT analysis.
   - **C.** completing a community risk hazard analysis.
   - **D.** crafting a strategic plan.

CHAPTER 9

Working in the Community

Fire Officer III

Knowledge Objectives

After studying this chapter, you should be able to:

- List the roles and responsibilities of a Fire Officer III in working with the community NFPA 6.3. (pp 209–211)
- Discuss the roles stakeholders play in a community and in relationships with the fire department NFPA 6.3. (pp 211–213)
- Discuss the chief officer's role in public education NFPA 6.3 NFPA 6.5.2. (pp 213–217)
- Explain how to reduce risk within the community NFPA 6.3 NFPA 6.3.1. (p 220)

Skills Objectives

After studying this chapter, you should be able to:

- Perform the roles and responsibilities of a Fire Officer III in terms of a community's needs NFPA 6.3. (pp 209–211)
- Communicate effectively with stakeholders NFPA 6.3. (pp 211–213)
- Employ public education effectively NFPA 6.3 NFPA 6.5.2. (pp 213–217)
- Reduce risk to the community NFPA 6.3 NFPA 6.3.1. (p 220)

Fire Officer IV

Knowledge Objectives

After studying this chapter, you should be able to:

- List the roles and responsibilities of a Fire Officer IV in working with the community NFPA 7.3 NFPA 7.3.1. (pp 221–222)
- Discuss the Fire Officer IV's stakeholder relationships NFPA 7.3 NFPA 7.3.1. (pp 222–225)
- Discuss the benefits of professional networking NFPA 7.3 NFPA 7.3.1. (pp 225–226)

Skills Objectives

After studying this chapter, you should be able to:

- Perform the roles and responsibilities of a Fire Officer IV in terms of a community's needs NFPA 7.3 NFPA 7.3.1. (pp 221–222)
- Communicate effectively with stakeholders NFPA 7.3 NFPA 7.3.1. (pp 222–225)
- Network with stakeholders and other groups NFPA 7.3 NFPA 7.3.1. (pp 225–226)

As you are about to sit down with your chief administrative official for a weekly update meeting, you receive a text message from your administrative assistant telling you that a local neighborhood association board member is waiting in your office to talk to you. The assistant has not been told the nature of the board member's business, but she reports that he does seem upset. You ask your assistant if she is able to reschedule his visit for later in day when you return to the office.

Your municipality's administrative official begins your weekly meeting by telling you that your annual evaluation is due. She presents you with her evaluation in which she identifies her concern for your lack of relations with community stakeholders. She suggests that you make a list of community stakeholders and develop actions and activities that will enable you to improve these relationships.

1. Which community leaders, both formal and informal, can you identify as community stakeholders?
2. What types of activities can you develop that could allow you to develop stronger, positive relationships with the identified stakeholders?
3. Why is working with these stakeholders important?
4. Is the chief officer the only department member who should be involved in improving relationships with community stakeholders?

Introduction

Career, combination, or volunteer; full- or part-time; in counties, cities, towns, villages, districts, and other municipalities large and small, chief officers and their subordinate officers are always public figures and have a number of vital responsibilities in their communities. This chapter identifies some common roles that fire officers are often expected to play; it also describes the interaction between the chief officer and key community stakeholders, individuals, and groups who can heavily influence the fire department's mission. Navigating the many interactions and intertwined relationships between the community and the department representatives, specifically the chief officer, is a necessary skill if the department hopes to gain community support of its operations.

Fire Officer III

Fire Officer III Roles and Responsibilities

A chief officer's specific community roles and responsibilities, and their relative importance, vary based on the type of department, its legal authority, its organizational structure, and the local political environment. Sometimes these roles and responsibilities are explicitly described in charters, articles of incorporation, local ordinances, job descriptions, or other official documents. They can be encouraged by administrative officials or elected leaders or, in some cases, even prohibited by these same officials. In other cases, chief officers' roles and responsibilities arise from history, tradition, or past practice, with each successive chief officer simply following his or her predecessors without much thought as to why or how these actions developed or considering whether practices could be improved or eliminated. Just because a responsibility is not formally articulated in a legal document does not mean it is not important or, for that matter, that it *is* important. Rather, each responsibility should be reviewed periodically to ensure it is being approached in a manner that gains the greatest benefit for the department's mission. One thing we do know is that chief officers can often have a great impact on their communities through informal stakeholder interactions.

Depending on their fire department's size and organizational structure, chief officers may or may not be directly engaged, on a daily basis, with fulfilling all the roles and responsibilities described in this chapter. In larger departments, the fire chief may have a number of subordinate officers whose primary duties encompass many of these roles. In smaller departments, the chief officer might have to interact with the community directly to fulfill all of his or her responsibilities. Whatever the specific daily arrangement, it is important to remember that although authority may be delegated, the ultimate responsibility for meeting stakeholders' expectations always rests with the chief officer. Keeping this in mind, there are many occasions in which the community expects to see, hear, and interact with its fire department's chief officer; in these cases, there can be no substitutes.

Administrator

The fire chief's involvement in the business side of the fire department can range from serving as its chief executive officer, responsible for all facets of the department's operations and administration, to the chief operating officer, with primary responsibility for overseeing the fire and emergency services delivery component. An elected or appointed fire commissioner, board of directors, or president may handle the business aspects of running the organization. Regardless of the specific arrangement, community members generally expect the chief officer to be familiar with the administrative aspects of managing their fire department. Being an effective administrator means ensuring that the proper systems are in place that put the department in the best position to be successful. Developing, implementing, and evaluating systems that include employee recruitment, training, discipline, maintenance of department assets, fiscal responsibility, and both internal and external communications are administrative responsibilities. Chief officers must realize that CEOs and COOs are ultimately responsible for the success or failure of the organization. Excuses are not acceptable when the department fails to meet its mission of protecting the public's assets or if it is unable to operate within the financial constraints of its approved budget.

Subject Matter Expert

Whether elected by the fire department's membership; appointed by a local board, council, or commission; or serving at the pleasure of a mayor or city manager, chief officers are generally expected to be the principal fire protection subject matter experts for policy makers, elected officials, local government administrators, business owners, and citizens throughout their communities. The role of subject matter expert is a critical one, and it demands that fire officers remain current with all aspects of their department's mission.

Because it is extremely difficult for an individual officer to develop specialized expertise across every area of a department's operation and administration—especially in all-hazards departments that deliver fire prevention and suppression, emergency medical services (EMS), hazardous materials, and technical rescue services—successful chief officers are usually generalists and focus on cultivating teams of specialists within their organization and community.

The chief officer's primary duty as subject matter expert is providing public guidance on community needs related to fire protection and emergency services, along with describing the resources available to meet those needs. In some cases, fire departments are sufficiently resourced to address the community's daily risk of fire and other emergencies, short of an overwhelming disaster. In these instances, chief officers often find themselves explaining why the current level of resources provided to the department should be maintained to support current service levels. Sometimes community leaders view the chief officer as pessimistic—always crying that the sky is falling—as the chief officer attempts to hang on to vital resources during austere times with various "what if" scenarios.

In jurisdictions where the available fire department resources are not sufficient to meet service demands safely, chief officers become salespeople, soliciting support for expanded resources as they attempt to identify disparities between community needs and resource availability, along with the consequences of not filling those gaps. Simply sounding the alarm is not enough; chief officers must be able to create realistic plans for addressing limitations in their departments' capabilities. In some cases, this might mean not providing certain services or developing alternative funding sources such as EMS billing, cost recovery (e.g., malfunctioning alarm fees, utility responses, fire incidents), technical rescue and/or hazardous materials response fees, and even selling subscriptions for structural firefighting services. (See additional information in the "Budget and Finance Issues" chapter.) In others, it may be an agreement with a neighboring department to provide the service for a fee or other arrangement, including a functional or total consolidation of service.

Citizens, working through their elected representatives, make the ultimate decision about what to expect from their fire department. In debating the decisions on fire department service, however, elected officials—and more often, citizens—are sometimes led by relatively uninformed people and they may make decisions from positions of emotion or fear. In these cases, they cannot make intelligent, effective decisions without substantial input and credible subject matter expertise provided by chief officers.

Consultant

Closely related to the role of subject matter expert is that of principal consultant for the community on matters of fire protection and emergency response. Chief officers are expected to provide proposals for enhancing their department's ability to address community hazards. Beyond maintaining existing services, chief officers must be attuned to changes in their communities that have the potential to affect service provision, such as reviewing the impact of a newly proposed high-hazard industrial facility. In these cases, fire chiefs should provide substantial input on development and zoning proposals, where applicable, to ensure that services keep pace with community demand characteristics. It is also important for the chief officer to be thoroughly familiar with the municipalities' adopted fire codes and flexible with their application, while remaining within the intent of the code, to assist in locating new facilities. The chief officer would be best viewed as an ally of potential developers: the person who is looking out for the safety of their employees and their long-term continuity of business.

Incident Manager

While being an administrator is an important role for the chief officer, the chief cannot forget his or her tactical responsibilities or allow himself or herself to fall behind in tactical knowledge. Depending on the size and type of department, its geographic service area, and its organizational structure, some chief officers spend a great deal of time actually responding to emergency incidents, assuming the role of incident commander, and developing the tactical plan to mitigate the incident safely. In larger departments, routine daily operations and incident command responsibility are delegated to subordinate officers. When major incidents occur, however, communities often expect the chief officer to be present and involved at the emergency scene. Even when not on scene, chief officers will ultimately be held accountable for the performance of the department's employees. A chief officer may be expected to closely monitor all operations, to be available for response, or, at a minimum, to remain available for contact with on-scene resources. In other words, communities often expect their chief officer to be an employee around the clock.

Fire Marks

Rural/Metro Corporation

There are a number of privately owned and operated fire departments in the United States. Customers of these organizations often purchase subscriptions and are customers in the true sense of the word. Perhaps the most recognizable of these departments is the Rural/Metro Corporation, which provides fire protection in several U.S. communities. Rural/Metro provides EMS and/or fire services in 21 states, serving more than 690 communities (Rural/Metro).

As a public corporation listed on the NASDAQ stock exchange, Rural/Metro is primarily accountable to its investors, stockholders, and those who contract or subscribe to receive its services. In communities served by Rural/Metro, however, fire officers generally fill the same roles as their municipal counterparts.

■ Civic/Community Associations

Local civic/community associations and philanthropic groups are often powerful stakeholders in the community. Civic associations can be organized around a particular geographic area but may also represent groups of citizens with common or charitable interests. Chief officers who proactively cultivate relationships with civic and community associations often benefit from their members' support for the fire department's mission FIGURE 9-10. As partners in prevention, civic associations can also be advocates for community risk reduction, both before and after a fire.

■ Homeowners' Associations

Like some civic associations, homeowners' associations and community associations usually represent a given geographic area. These associations often interact with their local fire departments around issues such as community covenants and restrictions, fire code requirements, code enforcement matters, and resource deployment. Some homeowners contract with private providers or neighboring communities for fire protection and EMS; others have even formed their own fire departments.

© UPI Photo/Bill Greenblatt/Landov.

FIGURE 9-10 Fire departments can host events for civic groups at fire stations to help maintain connections with the community.

■ Special Interest Groups

There is almost no limit to the variety of special interest groups that can form in a community. Beyond those mentioned previously in this chapter, these include taxpayers' alliances, government oversight (watchdog) groups, environmental protection organizations, issue-based advisory councils, faith- and ethnic-based organizations, and labor unions or associations.

Networking

There are few truly novel challenges facing fire departments today. Networking is often the best way to identify potential solutions for problems that other departments have already successfully addressed. Regardless of their specific organizational positions, roles, and responsibilities, all chief officers benefit from professional networking within their local, regional, state, and national communities of practice. A chief officer who eschews participating in his or her regional and state chiefs' associations or committees risks becoming ill-informed or isolated when issues of mutual concern arise. Similarly, a chief officer who participates in developing state/national codes and standards can help influence best practices and, at a minimum, understand contemporary trends and issues.

Experienced chief officers know the value of providing subordinate officers with opportunities to interact with their counterparts in local partner agencies and organizations. Examples include the following:

- Attending police roll-call training
- Riding along with public works employees
- Conducting joint plan reviews with planning department officials
- Working through shared problems in emergency management exercises and real-world events
- Providing first aid, CPR, or other types of safety training to other municipal employees
- Holding regular informational sessions between organizations
- Holding regular training with other nontraditional organizations

The value of chief officers' local networking extends to attending civic association meetings and other stakeholder events (such as sporting contests, parades, block parties, and potluck dinners). It is rarely possible for a chief officer to personally attend everything that happens in even a small community. Some fire departments assign company or chief officers to work regularly with specific groups; these officers are an integral part of the community's conduit to the department.

Whether such arrangements are formal or informal, they are vitally important. At the scene of an incident or when dealing with a fast-moving public policy issue is not the best time or place to exchange business cards and start establishing a collegial working relationship. Furthermore, it is sometimes true that things the fire chief cannot accomplish at the upper levels of the organization can be swiftly handled by mid-level chiefs or company officers working with their colleagues in other agencies, departments, or stakeholder groups.

You Are the Chief Officer Summary

1. Which community leaders, both formal and informal, can you identify as community stakeholders?

Every community has its stakeholders. These individuals can become allies of the chief officer if he or she can identify them and find ways to interact with them on a personal and organizational level. A chief officer should look to form relationships with formal and informal leaders including:

- School district leaders
- Industrial and commercial business owners and managers
- Nonprofit group leaders
- Neighborhood groups and community activists
- Government officials within and outside their own community (local, state, and national levels)
- Housing associations
- Special interest groups

2. What types of activities can you develop that could allow you to develop stronger, positive relationships with the identified stakeholders?

Your first effort to build relationships with your community stakeholders is to identify ways in which you can utilize your resources to assist them and their organizations. For example, you may offer fire extinguisher training to employees of local businesses. Neighbor groups may wish to utilize your facility's meeting space for their meetings. Visiting schools, opening stations to touring children, providing apparatus displays at neighborhood block parties, and partnering with nonprofits on fund-raisers can place you—and more importantly, your department—in a positive light.

3. Why is working with these stakeholders important?

Working with community stakeholders puts you in front of the community's leaders. These relationships allow you to market your department's services and advocate for proper funding levels or expansion of service when needed. Building support throughout the community is important when dealing within the political environment.

4. Is the chief officer the only department member who should be involved in improving relationships with community stakeholders?

The simple answer is no. Everyone from the fire fighter to the fire chief should be involved within the community. This demonstrates a total commitment of the department to meeting the needs of the community. Many of the activities in which the department can become involved do not necessarily require the chief officer. Fire fighters and line officers can become excellent advocates of the department's mission.

Wrap-Up

Chief Concepts

- Navigating the many interactions and intertwined relationships between the community and the department representatives, specifically the chief officer, is a necessary skill if the department hopes to gain community support of its operations.
- A chief officer's specific community roles and responsibilities vary based on the type of department, its legal authority, its organizational structure, and the local political environment.
- Chief officers are generally expected to be the principal fire protection subject matter experts for policy makers, elected officials, local government administrators, and citizens throughout their communities.
- Chief officers are expected to provide proposals for enhancing their department's ability to address community hazards.
- A successful approach to advocacy is to remain politically savvy without becoming politically aligned or partisan.
- Chief officers are usually the primary spokespeople for their departments when communicating with stakeholders throughout their communities.
- The amount of complaint-driven activity is often influenced by how much, or how little, the fire department proactively cultivates positive relationships with its stakeholders.
- Chief officers must understand the political, economic, and social dynamics of their entire community.
- Networking is often the best way to identify potential solutions for problems that other departments have already successfully addressed.

Hot Terms

Advocate Vocal supporter of a cause.

Community A group of people linked by geographic proximity and/or a set of common interests.

Demographics The characteristics of the human population within a given community; for example, age, education level, or household income.

Geographic information system Equipment that analyzes and displays relationships between statistical data and geography.

Municipal Belonging to a unit of local government, including a city, county, town, township, district, authority, etc.

Not-for-profit An organization that measures its success in outcomes that are not reflected in bottom-line economic profits.

Stakeholders Organizations, groups, or individuals who can influence or be affected by the actions of another organization.

Walk the Talk

1. Working with internal stakeholders is important. You and your fellow department heads all compete for the same budget dollars. How does the chief officer fight for his or her share of the budget without alienating his or her peer group?
2. Obtain a copy of your municipality's disaster response plan and review it for its ability to respond to the various risks your community faces. Determine whether the plan is sufficient in its identification of the involvement, authority, and planning that your department will be asked to perform in the face of a major incident covered by the plan.
3. The text recommends that chief officers be politically savvy without becoming politically aligned or partisan. Who determines when an agency head is perceived as being partisan?
4. An international corporation has selected your community for an industrial processing facility. To win this decision, the mayor waived property tax payments for the first decade. The chief officer identified significant hazardous material risks within the facility. The workload generated by this facility will require creation of a dedicated special hazards company with extensive specialized training. The mayor says there is no money for new personnel, training, or specialized equipment. How can the chief officer engage the community to improve the department's ability to protect the industrial processing facility?
5. Geographic information and demographic data can describe and analyze the community down to a house-by-house level. Predominantly used to determine emergency service workload, how can a chief officer use this process to improve effectiveness in promoting the fire department mission to community decision makers?

CHIEF OFFICER *in action*

Like many municipal governments, your city is experiencing very tight fiscal times. Development has slowed and tax revenues are substantially reduced, while citizens demand further cuts and struggle to make ends meet. Although the elected officials and city management are sympathetic to your concerns about reducing the fire prevention budget, you are forced to eliminate overtime for night/weekend inspections of vulnerable occupancies such as nightclubs, restaurants, and other places of public assembly.

As you consider your options for providing this essential service, you think about how you might engage stakeholders in protecting their community. A week after the budget is adopted, you invite several groups to attend an open public meeting about how the fire department's service level will change due to the overtime cuts, including the chamber of commerce, small business association, Rotary club, Jaycees, police department, fire fighters' union, and several others.

After framing the situation and inviting participants' ideas, several groups step forward to partner with the fire department to create a voluntary inspection program informed by subject matter experts from the fire prevention bureau. The police chief offers the opportunity for fire department instructors to deliver roll-call training, enabling officers to spot the top ten most frequent violations while on routine patrol. Suspected violations will be referred to the fire department for follow-up the next business day, or sooner if the severity warrants immediate action. Recognizing the importance of proactive fire prevention inspections for helping to keep fire fighters safe, the fire fighters' union suggests training and certifying some of its members as fire inspectors so they can address complaints/violations while working night and weekend shifts.

1. Which of the following is not among the fire officer's roles and responsibilities?
   - **A.** Advocate
   - **B.** Spokesperson
   - **C.** Lobbyist
   - **D.** Subject matter expert
2. The daily authority to address certain fire department roles may be delegated to multiple fire officers, but ultimate responsibility for the fire department's administration and management rests with:
   - **A.** the captains.
   - **B.** the contractors.
   - **C.** the fire chief.
   - **D.** nobody; it's shared among various stakeholders.
3. Customer service is not important for municipal fire departments because they are government agencies.
   - **A.** True
   - **B.** False
4. Which of the following is not a key stakeholder in a local fire department?
   - **A.** Civic association
   - **B.** Chamber of commerce
   - **C.** State fire chiefs association
   - **D.** Police departmentPersonal and Professional Development

CHAPTER 10

Managing the Code Enforcement Process

Fire Officer III

Knowledge Objectives

After studying this chapter, you should be able to:

- Explain the history of fire codes NFPA 6.5. (pp 231–233)
- Explain how a fire safety bureau is developed NFPA 6.5 NFPA 6.5.1 NFPA 6.5.2. (pp 233–234)
- Discuss the code adoption process NFPA 6.5 NFPA 6.5.1 NFPA 6.5.2. (pp 234–235)
- Describe the steps in code enforcement NFPA 6.5 NFPA 6.5.1 NFPA 6.5.2. (pp 235–236)
- Identify and describe the budget impacts of code enforcement NFPA 6.5 NFPA 6.5.1 NFPA 6.5.2. (pp 236–237)
- Describe the role of the chief officer in code enforcement NFPA 6.4.4 NFPA 6.4.5 NFPA 6.5 NFPA 6.5.1 NFPA 6.5.2. (pp 237–240, 243)
- Identify special considerations in code enforcement NFPA 6.5 NFPA 6.5.1 NFPA 6.5.2. (pp 243–244)
- Identify methods to address code enforcement NFPA 6.5 NFPA 6.5.1 NFPA 6.5.2. (pp 244–246)
- Identify and describe other fire prevention activities NFPA 6.4.4 NFPA 6.4.5 NFPA 6.5 NFPA 6.5.1 NFPA 6.5.2. (pp 246–249)

Skills Objectives

After studying this chapter, you should be able to:

- Develop a fire safety bureau NFPA 6.5 NFPA 6.5.1. (pp 233–234)
- Participate in the code adoption process NFPA 6.5 NFPA 6.5.1. (pp 234–235)
- Enforce fire codes NFPA 6.5 NFPA 6.5.1. (pp 237–240, 243)
- Participate in building plan review NFPA 6.5 NFPA 6.5.1. (pp 247–248)

Fire Officer IV

Knowledge Objectives

There are no Fire Officer IV knowledge objectives for this chapter.

Skills Objectives

There are no Fire Officer IV skills objectives for this chapter.

A major employer in your jurisdiction has been operating for over 20 years in the same building. Their operation involves the manufacture of automobile parts, including a paint line. As the chief officer you notice that the plant is due for its fire inspection.

After the fire inspection is complete, your fire safety specialist (FSS) asks to meet with you to discuss problems associated with the inspection. She advises you that the paint line does not meet code compliance because it was modified without a permit. Additionally, there were numerous other violations, including untested fire extinguishers, inoperable exit lights, hazardous wiring violations, and a disconnected supervisory fire alarm panel. During the exit interview, after completing the inspection, the FSS was told by the irate owner that making the cited corrections would force him out of business.

1. What are the parameters for an ethical and effective resolution of conflicting desires when solving a fire prevention issue?
2. Which other municipal officials should be notified about the issues raised by the fire inspection?
3. Which steps should be taken by the chief officer to assist the FSS with gaining compliance, minimizing political fallout, and calming the affected business owner?
4. Should the administrative or executive chief officer have the same professional credentials as the inspectors and investigators?
5. Using the concept of engineering–education–enforcement, what are the trigger points to move from education to enforcement?

Introduction

The challenge presented to the chief officer in terms of code enforcement is that the chief officer may be required to quantify a negative when justifying the addition or continuation of a program. For example, the chief officer is put in a position of saying, "Because we have an aggressive code enforcement program, there were fewer fires in the business community." Appointed and elected municipal officials are looking for hard data that can back up such statements when making budget decisions. Effecting behavioral changes and/or changes in attitude is not an easy task, because solutions to these problems are sometimes culturally developed. An additional problem for the chief officer is that it may take years or even generations to develop the proof that these programs are working and that they are cost effective.

It is the chief officer's responsibility to sell the philosophy that the easiest fire to fight is the one that never occurs. Sometimes this job of salesmanship is necessary inside the department as well as in the community. Fire fighters typically would rather see money spent to put more personnel on the rig than to add code enforcement positions. Again, it is the chief's job to convince the people in suppression that the inspectors are making the fire fighter's job, and citizens' lives, safer. This same line of reasoning can and should be made to the elected and appointed officials. When there are fewer or smaller fires, there is less potential for injuries to the fire fighters, and businesses that do not have fires can remain a viable asset in the community. A 2013 report issued by the National Fire Protection Association (NFPA) found that 29,760 (or 45.2 percent) of the injuries suffered by fire fighters occurred on the fire ground (NFPA 2015). Because the continuity of operations is important to any business, it makes sense that business owners would consider the overall safety of any community an important issue when choosing to locate a facility. The chief officer must be prepared to argue for maintaining a community's desirability by improving its safety quotient.

Fire Officer III

History of Codes

Although there are model fire codes published by organizations like the International Code Council and the NFPA, there is no single national fire code in the United States. A local government's authority to establish and regulate fire safety standards is regulated by the state.

In some states, those limits are loose, and states generally allow local governments to act as long as there is no state law prohibiting such actions and no conflict between what the local government wants to do and what the state does. (In the event of such conflict, the state usually wins.) In other states, those limits are very tight and local governments cannot act unless there is a specific state law authorizing the local government action. In some states, the regulations are a combination of loose and tight, depending on the regulation.

Historically, fire codes began as local documents. States viewed fire protection as a local matter and made no attempt to regulate fire code development or enforcement. The result was a patchwork quilt of codes and regulations, and owners of buildings in multiple cities within a state often were not enthusiastic about the wide range of compliance standards. Adding to the building-owner confusion was a growing divergence between building codes and fire codes.

Some states responded by adopting a mandatory statewide fire code with no authorization of local standards that were either less or more stringent than the state code. These states adopted what are called mini-maxi fire codes—in which the state provisions provided both the minimum and the maximum standards. Other states adopted a minimum (mini) code; local governments could adopt more stringent standards, but not less stringent standards. A handful of states did not adopt a statewide code, but the legislature reserved the right to review and approve each local code (or to create a state agency or committee to conduct those reviews). Finally, a few states adopted statewide building codes and left fire codes to local governments, as long as the local fire code did not conflict with the state building code.

Generally, fire code enforcement is seen as a local-government responsibility. Some states maintain a state fire inspection staff to inspect state or high-hazard facilities; others rely entirely on local-government inspectors. Similarly, some states establish training, certification, and continuing education standards for local inspectors, while others rely on local governments to define their own training, certification, and continuing education requirements.

At this point, chief officers may wonder about the federal government's involvement in the development of America's fire protection system. When it comes to basic fire service delivery and the development and enforcement of fire codes, the federal government's involvement has been marginal. Congress, the president, and the U.S. Supreme Court were predated by the 13 original states, and the participants in drafting the U.S. Constitution thought about fire protection as a local issue.

It took a combination of the English legal system (the forebear of the American legal system), the evolution of state law, and the adoption of the U.S. Constitution (with the accompanying rise of federal law) to yield the legal system within which today's chief officer works.

Fire codes are typically developed based on the experience gained from fires that have either caused a massive financial loss or a significant loss of life. As early as the 1st century AD, after the Great Fire of Rome, the Roman Emperor Nero developed fire codes to prevent a recurrence. He ordered the widening of streets and the separation of buildings, limited the use of combustible construction materials, and his fire fighters (known as "Vigiles") were able to enforce a fire code. Individuals found to be in violation of the fire code were typically ordered to be beaten.

It is unfortunate that major code changes frequently are instituted only when the need is proven through a loss of life. We must remember that fire codes are established to protect the lives of the citizens who occupy buildings and of the responders who enter those burning buildings for rescue and fire suppression.

Fire Marks

Events Changing Codes

Events that caused changes in fire and building codes in the United States include the following:

- In 1631, Boston became the first city in the American colonies to introduce a fire code. The new code outlawed wooden chimneys and thatched roofs because these were the most common causes of fire.
- In 1648, New Amsterdam (now New York City) established a program utilizing fire wardens, making it the first community in the country to have an organized fire inspection/prevention program. These inspectors checked for proper cleaning and installation of chimneys.
- The Great Chicago Fire of 1871, which killed at least 250 people and destroyed at least a third of the buildings in Chicago, surprisingly did not result in changes in the fire code. The city did enforce existing building codes for structures that were reconstructed, however, and Fire Prevention Week marks the anniversary of this fire.
- During the period of the Great Chicago Fire, John Damrell was elected chief engineer of the Boston Fire Department. He became a successful advocate for improved building and fire codes and the founder and first president of the National Association of Fire Engineers, now known as the International Association of Fire Chiefs.
- In a meeting on November 6, 1896, at the offices of the New York Board of Fire Underwriters, a group representing fire insurance organizations reviewed the articles for a new association and a set of proposed rules effecting fire safety. Article No. 1 stated, "This organization shall be known as the National Fire Protection Association." Today the NFPA's mission is "to reduce the worldwide burden of fire and other hazards on the quality of life by providing and advocating consensus codes and standards, research, training, and education" (NFPA .org). It has developed and adopted over 300 codes and standards toward this end.
- On December 30, 1903, a fire occurred in Chicago's Iroquois Theater. Every seat was filled and hundreds of patrons filled the standing room areas at the back of the theater. Some patrons chose to sit in the aisles, blocking the exits. A fire started in scenery above the stage and quickly spread. Five hundred ninety people died because they were unable to exit. Code changes that resulted from this fire included a requirement for panic hardware, exiting pathways, exit doors, exit signs and markings, and maximum seating capacities.

Fire Marks

- In March of 1911, the Triangle Shirtwaist factory fire occurred in New York City. Triangle Shirtwaist Company was located on the 8th through the 10th floors of a 10-story building. A fire broke out on the 8th floor and quickly spread to the floors above. Nearly 150 workers died due to their inability to flee the flames because of locked and blocked exits and inadequate fire escapes. Resultant code changes included fire proofing, sprinkler systems, improved exiting from high rises, and development of NFPA 101, *Life Safety Code*.
- In April of 1940, the Rhythm Club in Natchez, Mississippi, burned, killing 241 people. The fire quickly engulfed the structure, enabled by the decorative Spanish moss draped over the rafters. Windows had been boarded to prevent outsiders from viewing or listening to the music. The occupants were trapped. Code changes that resulted include standards for the direction of door swing, the number of exits required, and regulations on interior finish.
- In November of 1942, the Cocoanut Grove nightclub in Boston burned, resulting in nearly 500 fatalities from the combination of a fast-moving fire and locked or blocked exits. Code changes that were influenced by this fire include outward-swinging doors, fire suppression systems, collapsible revolving doors, the number of exit doors, battery-operated emergency lighting, exit access width, limitations on construction materials and interior furnishings, and additions of codes to NFPA 101, *Life Safety Code*.
- In December of 1946, 120 people died in the Winecoff Hotel fire in Atlanta, Georgia. The fire covered the third through the fifth floors and is considered the worst U.S. hotel fire until that time. The resulting code changes include the location of fire exits, fire suppression systems, and alarm systems.
- In April 1949, a rapidly spreading fire in St Anthony's Hospital in Effingham, Illinois, claimed 74 lives. The code changes that occurred as a result included fire barriers, smoke barriers, and fire-resistant stairway enclosures.
- In August 1953, fire broke out in a General Motors Plant in Livonia, Michigan, and claimed the lives of six workers and caused $35,000,000 in damage. Code changes influenced by this fire include restrictions on roof tar buildup, separation of hazardous operations, sprinkler requirements in industrial buildings, fire coating for steel frame trusses, automatic fire doors, and development of NFPA 204, *Guide for Smoke and Heat Venting*.
- On December 1, 1958, a fire broke out in the basement of Our Lady of the Angels Catholic school in Chicago, killing 95 students and nuns. While in compliance with the code when it was built in 1910, the school was grandfathered and not compliant with the code in 1958. Changes that resulted were the requirement of fire alarms, automatic sprinkler systems, self-closing exit doors opening outward, window egress heights, one-hour fire resistance rated walls, dedicated emergency lighting, separation of heating devices, and fire doors at stairwells.
- In May of 1977, a fire occurred in the Beverly Hills Supper Club that resulted in 167 fatalities. As a result of this fire, the use of aluminum electrical wire was banned, and public assembly buildings with a capacity of 301 or more were required to have sprinklers.
- The International Code Council (ICC) was established in 1994 as a nonprofit organization dedicated to developing a single set of national model codes. The founders of the ICC are Building Officials and Code Administrators International, Inc., the International Conference of Building Officials, and the Southern Building Code Congress International, Inc. The nation's three model code groups responded by creating the ICC and by developing codes without regional limitations–the international codes. As part of their efforts, they have established many subject-specific codes, including the *International Fire Code*.
- In February of 2003, a fire occurred in The Station nightclub in West Warwick, Rhode Island, killing 100 people. As a result of this fire, NFPA 101 was changed to lower the threshold for sprinkler requirement in nightclubs to 50 occupants for new clubs and 100 occupants in existing clubs.

Developing a Fire Safety Bureau

If a fire department's administration intends to have personnel regularly assigned to the responsibilities of plan review, fire inspection, code enforcement, and/or public education, the first step is to create a bureau, department, or division that will implement the program within the department. The *International Fire Code* Section 103 addresses the creation of a department of fire prevention in subsection 103.1 General: "The department of fire prevention is established within the jurisdiction under the direction of the fire code official. The function of the department shall be the implementation, administration, and enforcement of the provisions of this code" (International Code Council 2015, §103.1, 6).

Some communities may find the use of the word *department* confusing because it appears to create another entity that is equal to the fire department rather than a part of the department. Use of the term *bureau*, *branch*, or *division* eliminates this confusion and identifies a direct line of responsibility from the chief officer to the division or bureau chief officer. Choice among the terms *department*, *branch*, *division*, or *bureau* is dependent on the terminology used in the community. (For simplicity's sake, the term *bureau* will be used in this chapter.) In smaller organizations, there may not be an additional chief officer solely responsible for supervising these activities; these duties may be a part of the chief officer's duties, or personnel may report directly to the fire chief.

Another aspect of naming an organizational element is whether to call it the fire prevention bureau or something less intimidating, such as the fire safety bureau. In the future, small issues that are decided at the time the bureau is created will affect the fire department's approach to the community. The choice of a name will set the tone for future interactions with building occupants and owners. If the community is one in which the bureau will be taking an aggressive enforcement stance, then *prevention bureau* may be the right choice. On the other hand, if the community is looking to obtain voluntary compliance, then *safety bureau* may be preferred.

Additionally, the term used to identify the people who are in the field or on the phone talking to residents and business owners has the same effect. Some administrations prefer to send a fire safety specialist (FSS) to do a safety audit for safety concerns, rather than a fire inspector to do an inspection looking for code violations. When the FSS leaves, he or she will provide a list of safety concerns that need to be corrected. When the inspector leaves, he or she will provide a list of violations that must be corrected. Both individuals will cite the code for the justification of their findings. The FSS or inspector will return for a follow-up visit with the expectation that the problem will be corrected. If the problem is not corrected, either will increase his or her insistence that the problem be corrected. (The abbreviation FSS will be used in this chapter to identify the person making code enforcement visits.)

Adoption of Fire Codes

The FSS relies on an adopted fire code in order to ensure public safety within the community. Whether the code is adopted on a local or state level, it provides the FSS with the legal authority to mitigate the potential hazards from fire. Fire codes are also sometimes referred to as maintenance codes because they typically apply to existing buildings with new construction completed under the control of an adopted building code and enforced by the building official. Model fire codes are typically developed through a consensus process on a national level for adoption by state and local officials.

Nationally Adopted Fire Codes

Codes are intended to provide safeguards for the people who occupy a structure by developing standards for such things as access, egress, separations, and fire suppression systems. By adopting a nationally recognized code, the chief officer's community protects itself as well. A nationally adopted code has the weight of the organization's experience, and the members know that they are enforcing a legally vetted document. Over the years, there have been sundry developers of national fire codes, but the ICC and the NFPA publish the two prevailing codes.

Both the NFPA and the ICC update their fire codes on a three-year cycle. Some recent efforts have been made to force a change in the code adoption cycle from the established three years to six years. One concern with this move is the potential to delay needed code changes brought about by advancements in fire protection technology, building materials, and methods.

International Code Council

The ICC was created in 1994 as a nonprofit organization with the mission of developing a single set of comprehensive and coordinated model codes. It uses the governmental consensus process for code development. Only members representing governmental entities may vote on the approval of a code. The ICC comprises the three largest construction code organizations: Building Officials and Code Administrators International, Inc., International Conference of Building Officials, and Southern Building Code Congress International, Inc. Originally, these organizations developed individual sets of codes, but in 1994 they formed the ICC and began developing codes that were national rather than regional.

National Fire Protection Association

The NFPA is a nonprofit organization that anyone may join. Joining NFPA as a dues-paying member allows one to vote at the annual technical session and consider challenges to the actions of the technical committees—the first step in the appeal process. Members of the technical committee who write the codes or standards are not required to be NFPA members. Technical committee members are volunteers from various industries, and their participation is balanced so that none of the nine industry classifications can comprise greater than one third of the committee membership. Technical committee members must reach a consensus when developing NFPA documents. The NFPA's mission is to reduce the burden of fire and related hazards on the quality of life. The NFPA's technical committees have developed hundreds of codes relevant to fire safety, many of which have been adopted by reference in other fire codes. The following two of the codes developed by the NFPA are intended to enforce fire safety in general:

1. NFPA 1, *Fire Code*, covers the inspection of buildings, processes, equipment, systems, and fire and related life safety in existing occupancies, as well as the design and construction of new buildings, remodeling, or additions to existing buildings.
2. NFPA 101, *Life Safety Code*, addresses features to minimize the danger to people in a structure from the effects of fire, smoke, heat, and toxic gases created during a fire. It does not address general fire prevention or building construction features that are included in fire prevention and building codes.

NFPA 70, *National Electrical Code*, is another widely adopted NFPA document that plays a big role in how buildings are made fire-safe. Chief officers should be familiar with this code as well.

State Code Adoption

Some states adopt a code that is enforced by local municipalities. For example, the Utah State Legislature adopts the *International*

be agreed upon that only the chief officer or fire chief may approve code interpretations regarding intent.

Chief officers may also take a more active role in field inspections when issues become more complex or politically motivated. In these cases the chief officer serves as a shield to the FSS and is responsible for keeping the elected officials or chief administrative officer apprised of the status of the inspection and compliance process.

Inspection Scheduling

Inspections may either be scheduled or unannounced. One of the drawbacks to scheduling an inspection is that the owner/occupant may go through the building and correct all of the violations that he or she is aware of in an effort to pacify the fire department. This hardly seems like a drawback in that compliance is the intention of the visit. In fact, it does make the FSS's job much easier. The FSS should schedule an inspection for a large facility where it may take all of a day or more to thoroughly review the operation **FIGURE 10-4**. The FSS should schedule school inspections to avoid making visits on exam days, and he or she should schedule busy restaurant and store visits to avoid being there during their peak times. However, the FSS should also make unscheduled walk-through visits during those busy times to check for overcrowding and exit access.

Inspection schedules may also be influenced by special events or holidays. For example, because the day after Thanksgiving is one of the busiest shopping days, it is expected that retail establishments will see some of their highest occupant loads. With stock levels high due to the expected holiday rush, basic fire safety inspections (e.g., checking for blocked exit paths) are recommended.

System Testing

If a fire protection system is being tested, it is required in most cases that a fire department representative be present during the testing **FIGURE 10-5**. This can cause the fire department employee to spend a considerable amount of time waiting and watching gauges. Some departments that are familiar with a company's quality of work will require notification of when testing will take place so they can stop by at that time to make sure the test is being run

FIGURE 10-4 Schedule inspections at large facilities where a thorough inspection will take several hours.

FIGURE 10-5 In most cases, a fire department representative must be present for system testing.

according to the standard. Ultimately it is the company that is signing the inspection/test form swearing that the test was performed per the standard and that the equipment passed. The FSS should never open or close valves when testing is being performed; the owner's representative should do so. If there is a problem, the owner is responsible, not the FSS.

The Impact of Equivalencies

Sometimes granting an equivalency is necessary, but not if it sacrifices safety. If asked to grant an equivalency, the goal must be to meet the intent of the code. From a practical point of view, the code cannot be written to address every situation. For example, a code may say that every first floor exterior wall must have an apparatus access road to within 150 feet (45.7 meters) of all portions of that wall. The primary reason for this requirement is to allow the fire department to readily reach all areas of the structure with an adequate fire stream. Topography at the proposed building site may prohibit this from occurring on a portion of the B side and most of the C side of the building, in which case a strict enforcement of the code would prohibit the structure from being built. A possible compromise is to require an automatic sprinkler system throughout the structure, which ultimately results in a safer building. Remember: never grant an equivalency that diminishes safety!

■ Plan Review Process

A development review meeting (usually used for large or complex projects) with the property owner, his or her professional representatives, and the municipality's staff allows everyone an opportunity to discuss the proposed project prior to the formal submission of plans **FIGURE 10-6**. By making the code requirements known from the beginning, the property owner can accurately evaluate what it will cost to finish the project. The intent of plan review is to determine whether the structure will meet the code. At least two plan reviews should be completed: a site plan to determine that

© Jones & Bartlett Learning. Photographed by Glen E. Ellman.

FIGURE 10-6 A fire service representative, the property owner, his or her professional representatives, and the municipality's staff should discuss the proposed project prior to the formal submission of plans.

there is adequate water supply for the needed fire flow and fire department access and a plan review of the structure to determine whether it complies with egress, separation, and suppression requirements. Developers will often require a certified engineer with appropriate specialties to review plans and stamp them as compliant with applicable codes, but the AHJ should not assume that this has been done and should instead be sure to check the compliance report provided by the engineering firm. It is imperative that the fire department keeps a record of the progress of the construction and ensures that all access roads, hydrants, and other exterior requirements are completed as per the approved plan.

The Appeals Process

The code provides the owner/occupant—who is required by the bureau to make some effort to achieve code compliance—the right to appeal the order or decision of the bureau. This appeal is submitted to a board comprising individuals who, through experience, education, and training, are deemed to be qualified to render a decision. The board may rule on the interpretation of the intent or provisions of the code by the bureau or a proposed equivalent remedy made by the owner/occupant. They may not waive any requirements of the code. If an equivalent proposal is offered by the owner and the chief officer is not comfortable approving it, both parties may agree that the best solution is to let the board of appeals rule regarding the appropriateness of the equivalent proposal.

The FSS must be prepared to explain the appeals process to an owner or occupant who is questioning his or her code decision. The FSS must remember that the appeals process is a part of the adopted code and utilization of the process is the right of the owner or occupant. They must not take an appeal personally as a negative reflection of their expertise. The FSS should also be prepared to assemble needed documentation for the appeals board hearing as well as ready to testify on behalf of the municipality.

Chief Officer Tip

Plan Review Process

It is important to recognize that the plan review process should not be completed in a vacuum. There are many components to completing a plan review that involve additional municipal staff besides the FSS. Public works will review water, sewer, site access, and in some cases electrical supply; law enforcement may look at traffic flow or security features; building officials will ensure compliance with building codes and also involve other trades, including mechanical, plumbing, and electrical reviews; and planning staff will address zoning and compliance with any site restrictions or landscaping/appearance issues. Because a plan review flows through many individuals and departments, it is important to maintain communications among all parties. Joint plan review meetings are a good way to avoid project developers getting conflicting information or attempting to play one reviewer against another. It is also advisable to appoint a single individual as the main point of contact for the developer, controlling the flow of information between the municipality and the developer.

Mitigating Fire-Ground Risks

The late author and expert in building construction, Francis Brannigan, referred to the building as our enemy. When fire fighters enter a burning building, it is only a matter of time until they put out the fire or the building collapses. The better the building is built and maintained, the better the chances are that the fire will be extinguished. One goal of the FSS should be making the fire fighters' safety a priority by ensuring that the code is being followed during construction. Fire fighters should be educated that the FSSs are making the building safer and less of an enemy by ensuring that the maintenance of the building is being done so that those systems that protected fire fighters originally will still do so after 20 years.

Improving Preincident Planning

The FSS should develop a floor plan of the buildings as he or she is reviewing them. His or her plan should become part of a preincident survey that is available to all of the fire officers to review with their companies. Other helpful information would include the following:

- Access points
- Emergency contact information
- Utility controls
- Construction type
- Needed fire flow based on a percentage of involvement

- Hazardous materials stored within the building or on site
- Personnel traps
- Exposures
- Fire suppression and alarm system type and location
- Water supply points

Some of the records management systems provide a preplan component that allows the FSS to upload the inspection information into a computer program, making information sharing significantly easier. The key to this information being of value is the ease with which it can be accessed and fire fighter familiarity with the forms.

Insurance Services Office Impacts

The Insurance Services Office (ISO) collects information on the ability of over 45,000 communities nationwide to provide fire protection and analyzes the data using the fire suppression rating schedule. Based on this survey, a public protection classification (PPC) from 1 to 10 is assigned to a community. Once a PPC has been assigned to an area, the insurance companies use that class to determine the rates they will charge for fire insurance in that area. Class 1 is the highest rating for property fire protection, and Class 10 does not meet ISO's minimum criteria. Credit is given for a department's inspection program in the area of training, where it is recognized for pre-fire planning inspections. These inspections are to be done by company personnel and should be completed under the general criteria provided in NFPA 1620, *Standard for Pre-Incident Planning*. The inspections are intended for fire fighter familiarization. Hazards that are not eminent should be passed to the bureau for follow-up. Eminent hazards need to be corrected at the time of the visit by the suppression crew. The maximum credit possible for this area is 12 points when inspections are completed on all buildings annually. Partial credit may be received when the frequency of inspection exceeds one year, with no credit received when inspections are completed more than five years apart.

You Are the Chief Officer Summary

1. What are the parameters for an ethical and effective resolution of conflicting desires when solving a fire prevention issue?

International Fire Code Section 104.1, "Authority" makes it clear that the code enforcement official has the right and responsibility to interpret and enforce specific requirements of the code. The fire marshal or designated representative does not have the authority to waive specific code requirements.

To resolve conflict, the owner or occupant must show how he or she will meet the code requirement by offering a performance-based alternative to the interpretation made by the code official.

The IAFC Code of Ethics provides descriptions of desired behavior, including "Avoid situations whereby our decisions or influence may have an impact on personal financial interests."

When in doubt, seek legal counsel.

2. Which other municipal officials should be notified about the issues raised by the fire inspection?

Because some of the issues identified deal with work completed without a permit, the building official should be notified. The building official, once brought up to speed, will identify which special inspectors will be needed: electrical, plumbing, mechanical, etc.

3. Which steps should be taken by the chief officer to assist the FSS with gaining compliance, minimizing political fallout, and calming the affected business owner?

First, the chief officer needs to apprise his or her chief administrative officer of the potential conflict between the business owner and the fire prevention bureau. A memo outlining the events, concerns, course of action, and potential solutions may be required to bring elected officials up to speed. This would be important if it is expected that the business owner will take his or her concerns directly to the community's elected leaders. In an attempt to identify flash points for the business owner and to become educated of the owner's specific concerns, the chief official may want to schedule a formal code compliance review meeting with all interested parties, including the following:

- Chief officer
- FSS
- Building official
- Trades inspectors
- Business owner
- Plant safety official
- Operations manager or any other owner representative as determined by the owner

4. Should the administrative or executive chief officer have the same professional credentials as the inspectors and investigators?

Yes, in organizations where the administrative or executive fire officer is expected to participate actively in criminal and civil court cases involving code enforcement. The chief officer is functioning as an active member of code enforcement.

No, if the primary role of the administrative or executive fire officer is to function as the administrative leader of a section of appropriately credentialed inspectors and investigators.

Consult municipal and state administrative case law for specifics impacting your jurisdiction.

5. Using the concept of engineering–education–enforcement, what are the trigger points to move from education to enforcement?

Enforcement ranges from fines to immediate evacuation of the structure. The first trigger point is when there is a condition that creates an immediate hazard to life and health. A malfunctioning processing machine that is filling a building with toxic fumes or a nightclub that is 300 percent above its maximum allowable occupancy, are examples of conditions requiring immediate enforcement.

The second trigger point is when the owner or occupant makes no effort to comply with the fire prevention code after a series of efforts by the AHJ to obtain code compliance through a written notice of violation with deadline.

Wrap-Up

Chief Concepts

- It is the chief officer's responsibility to sell the philosophy that the easiest fire to fight is the one that never occurs.
- A local government's authority to establish and regulate fire safety standards is regulated by the state.
- If a fire department's administration intends to have personnel regularly assigned to the responsibilities of plan review, fire inspection, code enforcement, and/or public education, the first step is to create a bureau, department, or division that will implement the program within the department.
- Codes do not automatically have the force of law; however, in many areas the adoption of a specific fire code is predetermined by law, either state or local.
- It is the chief officer's responsibility to oversee the development of a process to enforce the codes that have been adopted by the jurisdiction.
- The code enforcement process can provide a revenue stream generated through the issuance of permits, cost recovery ordinances, special inspection fees, plan review fees, and potentially from fines.
- The chief officer should be involved in plan review and available for direction and approval of code interpretations or compliance plans.
- Special code enforcement considerations include political positions, loss of business, life safety issues, and residential considerations.
- Ways to address code enforcement include a letter of concern, a written notice of violation, abating an eminent hazard, a stop work order, a fire watch, revoking an occupancy or fire prevention permit, or a court order.
- A development review meeting with the property owner, his or her professional representatives, and the municipality's staff allows everyone an opportunity to discuss a proposed project prior to the formal submission of plans.

Hot Terms

Administrative search warrant A document issued by a judge or magistrate upon application by an official with enforcement power of an administrative agency (e.g., code official, city attorney, fire chief, or authorized representative) to check for evidence of noncompliance as required or authorized by any municipal ordinance or regulation.

Authority having jurisdiction (AHJ) An organization, office, or individual responsible for enforcing the requirements of a code or standard or for approving equipment, materials, an installation, or a procedure.

Codes Standards that are an extensive compilation of provisions covering a broad subject matter or that are suitable for adoption into law independently of other codes and standards.

Fire watch A temporary assignment of qualified personnel to watch for and control fire hazards and notify the fire department if a fire occurs.

Letter of concern A letter the FSS provides to a building's owner/occupant that lists all the concerns or violations, where they are located, why they are a concern, and when they should be corrected.

Mini-maxi fire codes State-approved codes that adopt the minimum and maximum of local codes.

Written notice of violation A formal written notice served to an owner, operator, occupant, or other such person responsible for a fire code violation. There is potential for civil or criminal action to be taken if the violation is ignored.

Walk the Talk

1. Interview an administrative or executive fire officer who works in code enforcement. What are the top three issues for that organization?
2. Describe the code adoption process for your municipality. Interview several elected officials in your jurisdiction to identify their chief concerns when adopting new or updated fire safety codes.
3. Review your adopted fire code and explain how your code appeals process works. Identify community stakeholders that are needed to fill out your appeals board.
4. Interview two fire code officials from organizations that have adopted NFPA 1 and two that use the International Fire Code. Identify their respective reasons for why they chose to adopt their specific code.
5. Develop a simple checklist that can be used by on-duty fire crews to conduct fire safety inspections of business properties.
6. Imagine that fire safety provisions of the state's adopted building code have come under attack from building trades groups seeking to eliminate them from the building code. Their reason for this action revolves around the added cost of these fire safety features and their belief that the added cost will slow the progress made in new housing starts since the last general recession. Items specifically identified for removal include residential fire suppression, arc fault electrical breakers, and basement fire protection of truss floors. You have been asked by the state fire chief's association to identify stakeholders who might support the inclusion of these safety features in the state building codes and to develop a strategy to influence the approval process.

A developer has proposed a residential single-family housing project in a wooded lakeshore area containing rolling hills with some steep grades. The developer's proposal contains site features that appear not to meet the adopted fire code for access and egress. The developer has stated that complying with the code will make the project financially impossible and destroy the intended "feel" of the high-end development.

As the chief officer, you meet with your inspection staff to review the proposed plans. You ask them to review alternative solutions to the code issues that might allow the project to move forward and still provide the level of fire safety desired. The elected officials and the chief administrative officer are all in favor of the project and are pressing for you to resolve the identified issues.

1. With fire codes written by experts in their fields, is the AHJ able to modify provisions of the code?
   A. Yes, but only after making formal amendments to the code.
   B. No, the code requirements are set in stone once adopted by the elected body and cannot be changed.
   C. The AHJ may allow solutions to code issues that meet the intent of the code while maintaining the level of fire safety intended.
   D. If the developer can locate an example of how another community that has adopted the same fire code has dealt with the same issue, the solution developed by that other community will automatically apply to your situation.
2. You have been informed that the NFPA is considering changes to the standard in question that may provide you with alternatives for the developer. As the AHJ, at what point can you adopt the revised NFPA standard?
   A. As soon as you are aware of the proposed changes
   B. When comments on the proposed changes are requested .
   C. After the NFPA Technical Meeting in which the standard is discussed
   D. After the Standards Council issues the standard
3. Alternatives to fire access requirements might include which of the following?
   A. Addition of residential fire suppression to the building plans
   B. A signed hold harmless agreement for any damage or loss caused by the inability of the fire department to access the property during an emergency between the municipality, the developer, and the future property owners
   C. Installing a system of approved dry standpipes that allows the fire department to deliver water for fire operations quickly at the rear of structures not meeting the apparatus access requirements
   D. Both A and C may be considered.
4. The best procedure for resolving any code compliance decision of the code official is to:
   A. force the developer/owner to secure a vote of the elected official overturning your decision.
   B. submit a code change to the standards-making authority.
   C. allow the developer/owner to file an appeal of the AHJ's decision with the jurisdiction's code appeals board.
   D. require the developer to gather enough signatures on a petition asking for a public referendum on the issue.

CHAPTER

11

Emergency Management and Response Planning

Fire Officer III

Knowledge Objectives

After studying this chapter, you should be able to:

- Explain the fire service's role in emergency management and how it developed NFPA 6.6 NFPA 6.6.1 NFPA 6.6.2 NFPA 6.6.3 NFPA 6.8 NFPA 6.8.1. (pp 256–258)
- List and describe the basic responsibilities for emergency management programs NFPA 6.6 NFPA 6.6.2 NFPA 6.6.3 NFPA 6.7 NFPA 6.8 NFPA 6.8.1. (p 258)
- Describe how to use the National Incident Management System to plan for emergencies NFPA 6.6 NFPA 6.6.1 NFPA 6.6.2 NFPA 6.6.3 NFPA 6.8 NFPA 6.8.1. (pp 258–262)

Skills Objectives

After studying this chapter, you should be able to:

- Perform basic emergency management responsibilities NFPA 6.6 NFPA 6.6.2 NFPA 6.6.3 NFPA 6.7 NFPA 6.8 NFPA 6.8.1. (pp 256–258)
- Use the National Incident Management System to plan for emergencies NFPA 6.6 NFPA 6.6.1 NFPA 6.6.2 NFPA 6.6.3 NFPA 6.8 NFPA 6.8.1. (pp 258–262)

Fire Officer IV

Knowledge Objectives

After studying this chapter, you should be able to:

- Identify and describe the components of incident mitigation NFPA 7.6 NFPA 7.6.1 NFPA 7.6.2 NFPA 7.7.1. (pp 265–266)
- Describe the concept of preparedness and the preparedness cycle NFPA 7.6 NFPA 7.6.1 NFPA 7.6.2. (pp 267–270)
- Explain coordinated incident response and the assignment of emergency support functions NFPA 7.6 NFPA 7.6.1 NFPA 7.6.2. (pp 270–272)
- Identify and describe the steps of incident recovery NFPA 7.6 NFPA 7.6.2. (pp 272–273)

Skills Objectives

After studying this chapter, you should be able to:

- Coordinate an incident response NFPA 7.6 NFPA 7.6.1 NFPA 7.6.2. (pp 265–273)

The day starts out just as any other. The first sign of potential trouble comes when your city manager telephones and asks if you have seen the morning weather report. You have not. She advises you that the National Weather Service has issued an extreme risk of severe thunderstorms for your area with an expected arrival time of between 2:00 and 3:00 p.m. The conversation, although fairly short, leaves you with the understanding that the city manager wants to make sure that the fire department and the city are ready to respond to any potential outcome of the impending weather.

As you end the call, it dawns on you that no one has ever taken the time to develop those "what if" plans. Thoughts race through your head. Should you call in additional fire fighters? Were shelters identified and ready? With school in session and dismissal scheduled for 2:30 p.m., is it your responsibility to contact the district and to coordinate some type of emergency plan that up until now had not seemed very necessary? Would pushing the panic button look foolish if the storms never materialize?

Maybe attendance at the quarterly county emergency management meeting would have been a good idea after all. As the chief of the city's primary emergency response agency, you suddenly realize that planning goes hand in hand with response and even potential responses. It is a new day and you have just been placed on the front line of emergency preparedness, as storm clouds gather on the horizon. What you do next and how you perform might very well define your future role.

1. What can you accomplish by reviewing the community's past experience with severe weather?
2. What assistance should you ask of your municipality's agency administrator in planning for the potential threat?
3. As you prepare for the possibility of severe weather, what can you prepare to assist the local response agency leaders?
4. Which one of FEMA's four functional areas of responsibility for emergency management programs would involve reviewing the community's mutual aid response plans?

Introduction

The fire service has long viewed itself as the community's primary emergency response organization, with the key being the definition of response. To many residents, politicians, and others outside the organization, the fire service is viewed as singularly focused on fire and emergency medical services (EMS) based on the community's risk. Over time, new definitions of response have been established with specialized rescue, hazardous materials, expanded EMS transport, and other specialties added to the more traditional fire suppression role. Until recently, the idea of emergency management was not even recognized by many of those who drove the shiny red trucks. To them, the concept of emergency management was a group of individuals crowded into a small, lower-level room within the county building under the title of civil defense, a product of the Cold War. It was there that planning for the "big one," often assumed to be a nuclear holocaust, was the topic of most discussions. Some in the fire service may even remember the training offered on government-issued radiologic detectors—those yellow-colored devices with a red and blue-colored CD triangle logo, usually stored on some back room shelf.

Large-scale incidents, such as September 11, Hurricane Katrina, and Hurricane Sandy have placed much more attention on the redefined role of the emergency manager. Today, state-run emergency management programs with many new, diverse missions are in place to combat natural disasters, hazardous material incidents, and terrorism threats. Local emergency management programs are also encouraged to plan and prepare for these potential emergencies. Additionally, these state- and local-run programs are working more closely with numerous federal agencies and commissions. With this renewed sense of importance, the topic of emergency management has been reviewed, revisited, and documented to a much greater extent than previously. Today the line between the fire service and the traditional role of emergency management is much thinner. Communities across the nation, both big and small, are now either involved in emergency management or are looking for someone to take on emergency management tasks.

As communities look within to find the person with the right attitude and experience to fill the role of local emergency manager, the fire service, and consequently the chief officer, may find themselves with the opportunity to step up and demonstrate to the residents they are charged to protect that they are willing to accept this challenge. In examining the chief's role in emergency response, it is only fitting that the fire chief be viewed as the go-to local emergency manager. This chapter does not seek to serve as a one-stop source of information on the subject of emergency management; entire books are dedicated to this subject. Instead, it seeks to explore some basic examples of the responsibilities that confront the fire chief as a local emergency manager, the issues he or she may face, and the importance of that role to the future of the fire service.

Fire Officer III

Fire Service Role in Emergency Management

When determining how the responsibilities of emergency management fit into the traditional role of the fire service, it is perhaps best to explore the basics of emergency management responsibilities. The federal government, with the creation of the Federal Emergency Management Agency (FEMA) in 1979, has led the effort to coordinate emergency response to the nation's major disasters. President Jimmy Carter believed that the national response effort should be consolidated under one department and led by an individual reporting directly to the president. It was FEMA's first director, John Macy, who saw the similarities between planning for natural hazards and the older concept of civil defense. This led to the development of an all-hazard approach that would play into the partnership between the fire service and emergency management. After all, the fire service is called upon to respond to and assist in mitigating almost every emergency (barring law enforcement–specific responses, although even then the fire service may provide a supporting role) within a community.

After September 11, 2001, and the establishment of the Department of Homeland Security (DHS), FEMA's role was altered in that it now reports to DHS. The threat of terrorism shifted much of the focus away from the all-hazards approach. With the change, FEMA also lost its direct access to the president.

The fire service received its recognition as a leader in all-hazard response when, after Hurricane Katrina in 2005, President George W. Bush appointed former Fire Chief David Paulison as director of FEMA. Although attention to terrorism incidents was still a priority, planners at the federal, state, and local levels would be called upon to develop risk assessments and disaster contingency plans to prepare for catastrophic events including hurricanes, tornadoes, earthquakes, fires, and floods—incidents with a high probability of affecting various parts of the country on an annual basis. The goal was to create a seamless planning approach at all levels—local, state, and federal—that would result in a more successful, unified response during a time of need.

As chief officer, one should play a visible and primary role in the local emergency planning process. It is no longer deemed acceptable simply to sit back, wait for the disaster to occur, and then send in the troops—all the while expecting that success will just happen. There is no debate that it is the local fire service that will be called upon as the first responders to these local disasters. As much as FEMA and state emergency managers plan, prepare, and train, their resources are not immediate and in some cases may be days away. This leaves that initial and critical first response up to the fire department, the local responder. Just as the department is the primary response agency to that 2:00 a.m. structure fire or heart attack victim, it will also be the first to respond to a neighborhood devastated by a tornado or residents clinging to rooftops during a flood event. The difference is that, while the department is built to respond to those day-to-day emergency incidents, none of us is ever fully equipped to solely handle the response to major disaster events.

With proper emergency planning, local incident commanders (ICs) can reduce the time needed to realize the benefits of pre-identified resources and command assistance. It is these large-scale incidents where preparedness, mitigation, response, and recovery strategies developed well before the event have the biggest impact on how well the community survives the incident. Much like preplanning for an industrial fire, a fire chief would be well advised to participate in a lead role in his or her community's risk/hazard planning process. In some cases, this means getting involved at both the state and county levels because most disaster responses will involve resources from outside one's own jurisdiction. Additionally, the local emergency manager will need to understand the process of seeking state and federal aid through a maze of increasingly important disaster declarations. NFPA 1600, *Standard on Disaster/Emergency Management and Business Continuity Programs*, establishes a set of criteria for disaster and emergency management that can be of great assistance to the chief officer or other emergency planner.

Chief Officer Tip

Chief Officer Involvement

Chief officers should play a visible and primary role in their local emergency planning process. Do not expect success without active involvement. There is no debate that it is the local fire service that will be called upon as the first responders to local disasters.

The fire chief must also understand that he or she is not alone when it comes to emergency planning. It is in emergency planning that the chief can make great strides in serving as the catalyst in bringing together the other disciplines within the municipality. Since September 11, 2001, much attention within the emergency management realm has been given to the risk associated with terrorism. Law enforcement becomes the major player when dealing with the potential hazards associated with terrorist groups and threats and will take the lead in gathering intelligence data used to identify potential threats and develop appropriate response procedures.

The public works department, although sometimes forgotten as a first responder, can bring a multitude of resources and talents to disaster response efforts. One of the primary roles of the fire chief/emergency manager is to build an effective team with an eye on inclusion rather than exclusion. Excluding others from planning processes may lead to territorial fights, with various agencies falling into an unproductive protectionism mode.

As you read this, you may be asking how you can become more involved in your community's emergency management process. There are several steps that can place the fire chief in a better position to assume a leadership role for emergency management within his or her community, which are outlined in the following list.

1. Look to get involved at the county level. Find out how the county appoints individuals to its emergency operations center staff. Attend emergency management meetings and volunteer to work on local projects that involve planning, risk assessment, and exercising activities.
2. Educate yourself. Attend emergency management trainings and conferences. Share that knowledge with staff officers, fire fighters, and other municipal officials.
3. Examine the state of your own community with regard to emergency planning activities. Is there a formally appointed emergency manager position? If there is, what are the educational and experience requirements for the job and how would one meet them? If there is no such position, can you propose one or expand the fire chief's job description to include those duties? This will require convincing administration and elected officials regarding the need for someone to lead the emergency management direction of the municipality.
4. Is there a formal emergency plan for the municipality? When was the plan last updated? Can you volunteer to take responsibility for developing or updating the plan?
5. Look for ways to demonstrate knowledge of emergency management activities through inclusion of disaster response activities in the department's standard operating procedures.

Life Safety Initiatives

11. National standards for emergency response policies and procedures should be developed and championed.

Chief Officer Tip

Value of Response

While it is true that response to incidents can help prepare the chief officer for future incidents, they must also become thoroughly familiar with the ICS concepts of resource management necessary to mitigate the community impacts of a type 1 or 2 incident. Resource typing, coordination with emergency operations centers, use of general staff, and incident action planning are some of the items that are not usually associated with the day-to-day responses made by most chief officers, many of whom do not have the benefit of experience in acquiring or deploying resources for an expanding incident. In these cases, the chief officer will need to take advantage of training opportunities that include table tops and full-scale exercises in order to develop his or her disaster management knowledge and skills on a multiagency/jurisdictional level.

Fire Service Responses

All chief officers prepare their troops to respond to emergencies, and in doing so they prepare themselves for the command and control of these emergencies. These emergencies can be minor or routine in nature, can be a community or regional disaster, or can fall anywhere in between. In all cases, the fire department is called upon to respond and to mitigate the effects of the incident in a safe and effective manner.

Chief officers hone the incident management skills acquired through training in their many routine incident responses, sometimes on a daily basis. From the medical emergency to the second alarm structure fire, chief officers may be required to establish and fill the role of incident commander as well as work with any of the other command or general staff ICS positions. Today, ICS training is a mandatory item for any fire service officer meeting his or her federal NIMS requirements. In their role as incident commander, chief officers are required, among other things, to perform a scene size-up, establish objectives, develop strategy and tactics, and implement an action plan to resolve the problems identified through size-up. As the incident increases in complexity, the responsibilities of the incident commander will become more time consuming and resource driven, and there will be an increase in the communication needs of the incident. A greater emphasis on the functions of planning, resource management, and financial issues must be considered by the incident commander as the tactical needs of the incident grow and incident action plans go from being mental to written.

An important response function of the chief officer is to ensure that at the conclusion of a response incident operations are reviewed and evaluated so that any needed operational improvements can be implemented. The best way to identify needed improvements is through the process of a postincident analysis. A postincident analysis should include representatives from all operational agencies in an effort to receive the most accurate account of incident operations. If the critique is being conducted to review an exercise event, exercise evaluators and controllers must be in attendance in order to relate their observations. If timing of the critique means important information may be forgotten, ask agency officials to put their observations and experiences into a written report. Acquiring copies of local news reports can also be useful in reviewing a response. These documents can then be used to refresh memories at a later date.

As with any good critique, the group leader must make certain that the discussions are kept positive, with the intent of finding improvements and not pointing blame. The chief officer may serve as the moderator of the critique or, if intimately involved in the response, he or she may choose to find a neutral third party to moderate, thus preventing the appearance of bias. A participant should take notes during the critique, which can then be used to produce an after-action report. This report should be forwarded to all participants. The report should summarize major actions taken, along with outcomes and suggestions for improving operations noted. The report should also include those tasks that were handled properly to give credit to those who performed them. A follow-up meeting with key officials may be advisable to monitor assigned tasks for making improvements to the response system.

Being a chief officer in the fire service provides an excellent platform for one to assume a greater role in the community's emergency management responsibilities. The fact that chief officers are able to practice their emergency response, mitigation, and command skills on a regular basis through the various incident responses and the chief's knowledge of the ICS general staff positions (Operations, Planning, Logistics, and Finance/Administration) provides chief officers with a strong background for handling larger and more complex incidents. Through these daily responses, chief officers develop relationships with both internal and external stakeholders. Working with various agencies and mutual aid partners, chief officers can establish the network needed for dealing with incidents of a much greater scale. These high-risk/low-frequency events are more likely brought to a safe conclusion when they are guided by a competent and experienced commander. By gaining a thorough understanding of the total NIMS system, chief officers can be more effective during disaster operations.

In addition to their own responsibilities, chief officers must also be certain that their fire fighters and line officers are properly trained to handle the emergencies to which they will be dispatched. Ensuring a competent and safe response demands that chief officers determine the competence level of employees and their physical abilities to perform the necessary tasks. Training programs need to be evaluated to determine whether subject areas are adequately covering the operations, risks, and hazards that may be encountered on any given shift. This will require that chief officers make a commitment to a comprehensive training program that is sufficiently funded, adequately staffed, and documented with enforceable minimum training requirements.

Reviewing the current and potential operational demands of the department, identifying the hazards and risk of these responses that require employee training, and then providing appropriate and safe training do not complete the chief officer's response training responsibilities. The chief officer must also determine whether the training was successful. The testing of employees through simulated responses or by evaluating actual response outcomes through postincident analysis discussions can help quantify the fire fighter competency levels obtained and effectiveness of the department's training program.

Emergency Management Responsibilities

The fire service long ago recognized the value of prevention efforts in reducing death, injury, and property loss from a fire. The term *emergency management* might be taken by some to mean that we get involved in managing the emergency after it occurs. This thought could easily come from the false association with another fire service term, *incident management*. In the case of emergency management, the notion of getting involved in an emergency after the fact could not be more erroneous. In fact, the activities associated with emergency management might better be described as emergency planning and prevention because much more work needs to be devoted up front, before the event, in order to be successful.

Although the emergency manager may not be trained or geared to act as an IC, he or she could become the IC's best ally by lending the benefits of planning and resources to the incident response. When the chief officer becomes the emergency manager, the planning skills acquired for this new position—when combined with the management skills, knowledge of the National Incident Management System (NIMS) ICS system, and practical emergency response experience—can place the chief officer in a very strong position to lead the emergency management efforts of a community.

A chief officer is always involved with exploring the what-if potential of his or her protection area. In terms of a more traditional fire service, however, the focus is probably limited to more basic planning activities such as site-specific plans for hazardous materials incidents or evacuation plans for a senior high-rise or community hospital. Perhaps the responsibilities include an airport, rail yard, shipping port, or rural agricultural center. These are all hazards that were accepted with the position of fire chief. The chief officer has also acquired staff and directed training programs to prepare for these hazards while budgeting capital funding for needed apparatus and equipment. Does adding the responsibility of local emergency manager change the game? In simple terms it may not. The hazard analysis and risk assessment responsibilities that were undertaken for issues such as those listed above, combined with a vast knowledge of emergency response activities, should serve to prepare the fire chief in expanding his or her role to include that of local emergency manager.

FEMA is often considered the expert when it comes to activities surrounding the basic responsibilities of emergency management. Fire chiefs can obtain both information and training through FEMA and its partners on all aspects of emergency management.

To assist with the emergency management process, FEMA has developed four basic functional areas of responsibility for emergency management programs: mitigation, preparedness, response, and recovery. Reviewing and understanding these responsibilities will enable the fire chief to be better prepared for the disaster response when it does occur. Although they are unable to prevent an event from happening, these four functional areas can do the following:

- Better prepare a community for response to a potential emergency.
- Identify and build relationships with potential response partner agencies.
- Reduce the potential for damage from these catastrophic events if they do occur.
- Assist with a faster recovery effort.

NIMS Incident Typing

The chief officer should be very familiar with NIMS through its use on daily responses, but he or she should also explore the less often used functions that become paramount to success during major incidents. As a guide to determining the complexity of an incident and the use of some of the more advanced ICS components, FEMA has developed a system to categorize, or type, the level of emergency incident. As the level of the incident moves from a type 5 to a type 1, the level of complexity increases—as does the need for additional command support. FEMA's ICS-300 course includes a list of the types of incidents and some of their command needs TABLE 11-1.

It is the complexity of the incident that will drive the resource needs and increase the command components of the incident. As soon as local resources are overwhelmed and outside assistance is required, NIMS components such as the emergency operations center (EOC), area command (AC), incident action plans (IAPs), and/or incident management teams (IMTs) become important tools to the IC. These functions can help coordinate the response and ensure that the incident needs are being met.

Emergency Operations Center

The EOC can provide valuable assistance to the on-scene IC. Primarily a function of emergency management, the EOC provides resource management functions, assisting with formulation of policy and management of support services. The EOC support staff comprises various agency representatives who collectively bring a vast amount of knowledge to the support effort. The EOC is often structured using the emergency support functions as a guide for agency representatives. During disaster events or large-scale emergencies, the EOC is often called upon to locate resources from outside of the impact area and to assign them according to the incident priorities. What the EOC is *not* is the incident commander. Instead, the EOC interfaces with the on-scene incident command post as it supports the IC as needed.

When an emergency incident impacts a multijurisdictional area, the policy officials within the EOC may be called upon to establish resource and tactical priorities for the affected area. This is especially important when specialized resources are in short supply and great demand. EOC officials may also be utilized to collect damage assessments over the affected area. This facilitates preparing the documentation necessary for declaring a state of emergency on either a state or federal level. The management of support functions may also fall to EOC staff and partner agencies. For example, working with the Red Cross, emergency shelters may be established to house victims of emergency evacuations, freeing the incident commander from this responsibility.

Area Command

When multiple incidents are occurring within a defined area or jurisdiction, senior commanders may establish an

Table 11-1 FEMA Incident Types

Incident Type	Command Needs
Type 5	■ The incident can be handled with one or two single resources with up to six personnel. ■ Command and general staff positions (other than the incident commander) are not activated. ■ No written IAP is required. ■ The incident is contained within the first operational period and often within an hour to a few hours after resources arrive on scene. ■ Examples include a vehicle fire, an injured person, or a police traffic stop.
Type 4	■ Command staff and general staff functions are activated only if needed. ■ Several resources are required to mitigate the incident, including a task force or strike team. ■ The incident is usually limited to one operational period in the control phase. ■ The agency administrator may have briefings and ensure the complexity analysis and delegation of authority are updated. ■ No written IAP is required, but a documented operational briefing will be completed for all incoming resources. ■ The role of the agency administrator includes operational plans including objectives and priorities.
Type 3	■ When capabilities exceed initial attack, the appropriate ICS positions should be added to match the complexity of the incident. ■ Some or all of the command and general staff positions may be activated, as well as division/group supervisor and/or unit leader level positions. ■ A type 3 IMT or incident command organization manages initial action incidents with a significant number of resources, an extended attack incident until containment/control is achieved, or an expanding incident until transition to a type 1 or 2 team. ■ The incident may extend into multiple operational periods. ■ A written IAP may be required for each operational period.
Type 2	■ This type of incident extends beyond the capabilities for local control and is expected to go into multiple operational periods. A type 2 incident may require the response of resources out of area, including regional and/or national resources, to effectively manage the operations, command, and general staffing. ■ Most or all of the command and general staff positions are filled. ■ A written IAP is required for each operational period. ■ Many of the functional units are needed and staffed. • Operations personnel normally do not exceed 200 per operational period, and total incident personnel do not exceed 500 (guidelines only). • The agency administrator is responsible for the incident complexity analysis, agency administrator briefings, and the written delegation of authority.
Type 1	■ This type of incident is the most complex, requiring national resources to safely and effectively manage and operate. ■ All command and general staff positions are activated. ■ Operations personnel often exceed 500 per operational period and total personnel will usually exceed 1000. ■ Branches need to be established. ■ The agency administrator will have briefings and ensure that the complexity analysis and delegation of authority are updated. ■ Use of resource advisors at the incident base is recommended. ■ There is a high impact on the local jurisdiction, requiring additional staff for office administrative and support functions.

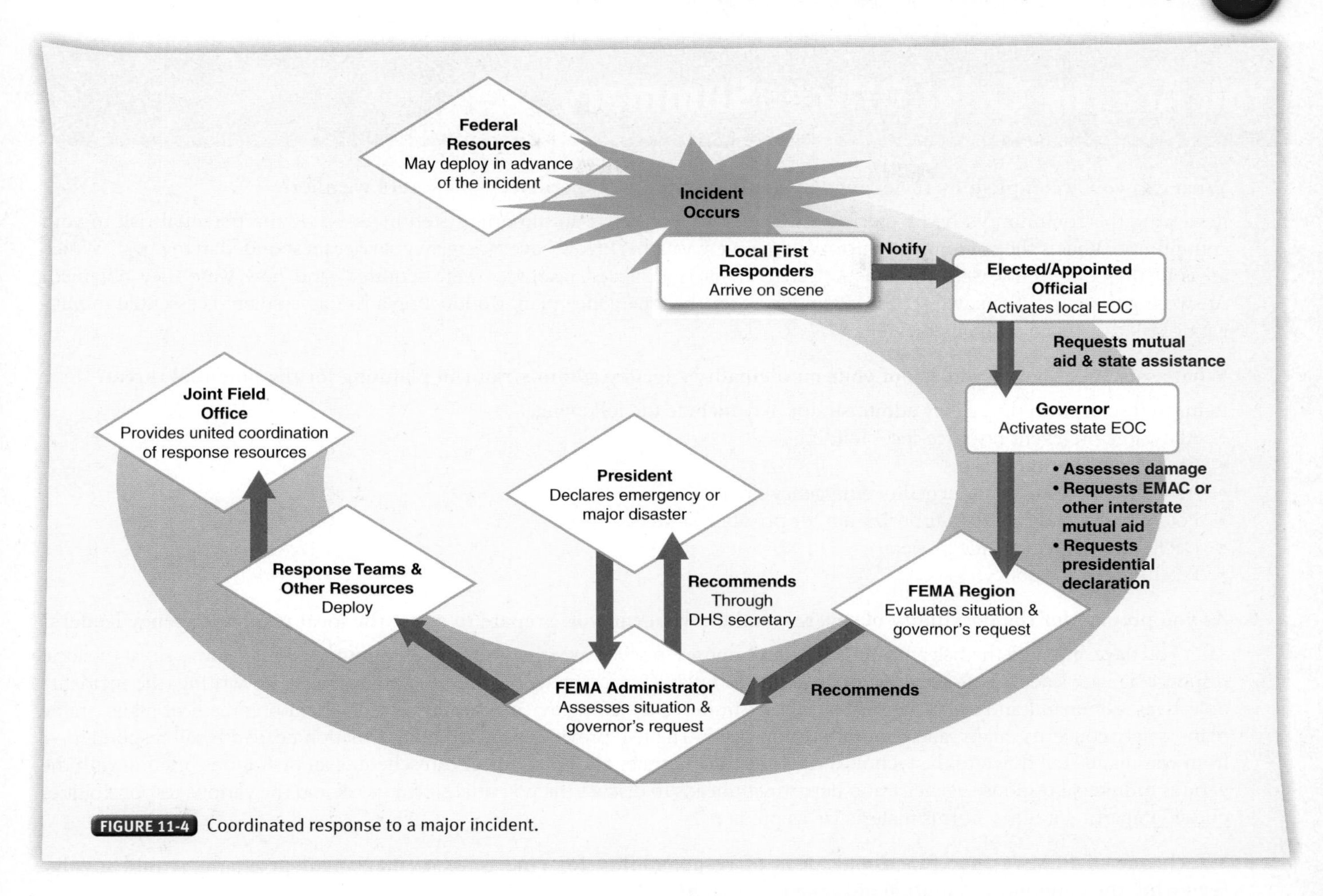

FIGURE 11-4 Coordinated response to a major incident.

Two FEMA publications—FEMA 325: *Public Assistance Debris Management Guide* and DAP 9523.13: *Debris Removal from Private Property*—can be referenced for questions relating to debris removal operations.

As part of the recovery efforts, ESF No. 14: Long-Term Community Recovery Annex, facilitates the long-term recovery planning process. Following state and local priorities, ESF No. 14 focuses efforts on permanent restoration of infrastructure, housing, and the local economy. Under the direction of FEMA, ESF No. 14 is supported by the Departments of Agriculture, Commerce, Homeland Security, Housing and Urban Development, and Treasury. The Small Business Administration can also provide assistance with loans to keep struggling businesses from closing.

You Are the Chief Officer Summary

1. **What can you accomplish by reviewing the community's past experience with severe weather?**

   Reviewing the community's past experience with severe weather is an important step in assessing the potential risk to your community. What is the community's history with severe weather? How have emergency services responded in the past? Which areas of the community were the most affected? What specialized resources were required, and how were they obtained? Answering these questions will set the groundwork for the emergency plan. Conducting a risk assessment is essential in mitigating risk associated with an emergency.

2. **What assistance should you ask of your municipality's agency administrator in planning for the potential threat?**

   Items to discuss with the agency administrator may include the following:
   - Allowable pre-event resource-level funding
   - Citizen notification process
   - Activation process of municipality emergency plan
   - Potential use and funding authorization for private resources
   - Declaration of emergency process
   - Potential curfew policy

3. **As you prepare for the possibility of severe weather, what can you prepare to assist the local response agency leaders?**

   After you have analyzed the risk and studied past responses to severe weather in your community, the best way to assist the local response agency leaders is to create an incident action plan (IAP). The IAP is used to communicate, in writing, the incident's objectives; command and organizational structure; resource assignments; and support needs, including medical plans, traffic plans, safety concerns, maps, and communication plans. The IAP provides a wealth of information needed by all responders—from command staff down to the technical level. As time permits, the chief officer can schedule an operations briefing with the various municipal response agencies and department heads to discuss the potential storm threat and the various responsibilities of each department if the storms materialize as predicted.

4. **Which one of FEMA's four functional areas of responsibility for emergency management programs would involve reviewing the community's mutual aid response plans?**

   Reviewing the community's mutual aid response plans would fall into the functional area of response. Response efforts include recognizing the resources available in your own community as well as those available through mutual aid agreements and other local, state, and federal organizations.

Wrap-Up

Chief Concepts

- Over time, new definitions of fire service response have been established with specialized rescue, hazardous materials, expanded EMS transport, and other specialties added to the more traditional fire suppression role.
- As chief officer, one should play a visible and primary role in the local emergency planning process. It is no longer deemed acceptable simply to sit back, wait for the disaster to occur, and then send in the troops—all the while expecting that success will just happen.
- The activities associated with emergency management might better be described as emergency planning and prevention because much more work needs to be devoted up front, before the event, in order to be successful.
- The chief officer should be very familiar with NIMS through its use on daily responses, but he or she should also explore the less often used functions that become paramount to success during major incidents.
- FEMA identifies six categories associated with mitigation efforts:
  - Prevention
  - Property protection
  - Public education and awareness
  - Natural resource protection
  - Emergency services
  - Structural projects
- FEMA's preparedness cycle calls on agencies to follow a continuous cycle of planning, organizing, and equipping responders; training; exercising plans and responses; and improving them through a formal evaluation process.
- Even major disaster responses begin with local first responders. The fire service has embraced this front-line role and has continually worked to improve its response posture. A good example of this is the expansion of mutual aid agreements into automatic aid, bringing outside resources to the scene on the original call, greatly enhancing on-scene capability.
- The fire service should be familiar with basic recovery efforts because they are often used, even on single-family residential structure fires. Fire service ICs know that when the fire is out the job is not done; efforts must begin to assist the newly homeless citizens.

Hot Terms

Area command (AC) A command center used to coordinate the management of multiple incidents that are themselves being managed by individual incident management organizations or for managing multiple incident management teams at a single large-scale event.

Emergency management The discipline charged with mitigation, preparedness, response, and recovery functions associated with disaster response.

Emergency operations center (EOC) A command center used to carry out the functions of emergency management during a large-scale event or disaster. Operating at the strategic level, EOCs assist on-scene ICs with logistical resources, policy, and other support functions.

Emergency support functions (ESFs) Functions that provide the structure for coordinating interagency support for a federal response to an incident.

Hazards Those items or events that have the potential to cause harm, damage, or otherwise negatively impact the community and its residents.

Incident action plan (IAP) A written plan of action outlining incident objectives, strategy, and tactics for a given operational period. The IAP also identifies the incident's command structure, safety concerns, and other logistical requirements.

Incident management team (IMT) A team of qualified and trained individuals deployed to assist local ICs with support, command, and/or control of an incident. IMTs are categorized by type; type 1 is national level; type 2 national/state; type 3 state/regional; and type 4/5 county/local.

Mitigation The efforts taken to reduce loss of life and property by lessening the impact of disasters when they do occur.

Mutual aid box alarm system (MABAS) A system of organizing resources for the purpose of responding to requests for mutual aid.

National Incident Management System (NIMS) A nationally recognized, comprehensive approach to incident management involving the following components: preparedness, communications and information management, resource management, command and management, and ongoing management and maintenance.

Operational period The period of incident operations covered by an incident action plan, usually between 4 and 24 hours, as determined by command.

Preparedness A continuous cycle of planning, organizing, training, equipping, exercising, evaluating, and taking corrective action to ensure effective coordination during incident response.

Preparedness cycle A continuous cycle of planning, organizing, and equipping responders; training; exercising plans and responses; and improving them through a formal evaluation process.

Recovery Efforts to return a community to normal after a major event or emergency. Recovery is the fourth function of emergency management.

Response A function of emergency management referring to the coordinated response of local, state, national, and federal resources and their coordination and command structures to incidents.

Walk the Talk

1. The last update to your emergency operations plan was completed over five years ago. What is the process required to complete an update of the EOP, and who should be involved?
2. Your municipality's dedicated emergency manager has just retired. Your agency administrator has told you that funding shortfalls have necessitated that the position be combined with that of an existing department head. Develop a proposal outlining why you, as the fire chief, should be appointed as the new emergency manager.
3. Identify ways in which you can expand your preparedness for disaster response through collaboration with outside agencies and officials.
4. List ideas for incorporating the training of municipal employees in disaster response activities into your existing department training program.
5. Develop a plan that uses the expertise of the fire department in the training of other municipal employees in disaster response operations.
6. Public health physicians warn that a pandemic would cause 25–35 percent of the workforce to be unable to come to work because they would be either sick or treating a family member who is sick. Some all-volunteer fire departments are staffed by personnel who work in emergency service or health care. During a pandemic, the able-bodied employees may be on a 12-hour on/12-hour off schedule. How can a volunteer agency respond to a pandemic?

CHIEF OFFICER *in action*

Budget cuts have led to the consolidation of job responsibilities for many of the city's department heads. You have just been assigned the responsibility for emergency management within your municipality. This is a position that you have prepared for, and you welcome the new challenges you will face. Previously, emergency management responsibilities were handled on a committee basis with the various department heads contributing as needed to the loose-knit process.

Your agency administrator has asked that you develop an emergency plan to handle potential catastrophic events. As the chief fire officer, you are keenly aware that your actions in this new position will be closely watched by the other members of the management team. Your first order of business will be to build support among these very leaders for a comprehensive emergency plan. Knowing that your success depends upon their cooperation in and acceptance of your planning process, you call your first meeting.

1. Working from scratch, with no formal plan to review or amend, what should your first action item be?
   A. Develop an IAP for a potential severe storm.
   B. Create a risk identification process.
   C. Establish policy for the operation of an area command.
   D. Prepare a budget request for upgrades to the city's emergency operations center.
2. Under Presidential Directive 5, what concepts should be incorporated into the municipality's first EOP?
   A. NIMS
   B. Governmental Accounting Standards Board (GASB) standards
   C. MABAS policies
   D. None of the above
3. The newly developed EOP outlines steps for the community's department heads to work together from co-located facilities at city hall to establish response priorities, resource allocations, overall objectives, release of public information, and city policy. This activity best describes an:
   A. EOC.
   B. IAP.
   C. AC.
   D. EOP.
4. Within your plan you provide for the training of civilians to assist local responders during disasters. Utilization of civilians for this activity is best described as a(n):
   A. IMT.
   B. EOC.
   C. CERT.
   D. DMAT.
5. If the effects upon your community of a disaster are severe and a presidential disaster declaration is needed, it would be requested by whom?
   A. A FEMA coordinator
   B. The Joint Field Office
   C. The person conducting emergency support function No. 4
   D. The governor

References and Additional Readings

Age Discrimination in Employment Act. 29 USCA § 623(f)(1) (1967).

Americans with Disabilities Act. 42 USCA § 12111 (West 2008).

Ancira Enterprises, Inc. v. Fischer. 178 S.W.3d (TX Appellate Court, June 16, 2005).

Aronson, J. Richard, and Eli Schwartz. *Management policies in local government finance*. Washington, DC: International City/County Management Association, 2007.

Autry, James F. *The servant leader.* Roseville, CA: Prima Publishing, 2001.

Barr, Robert C., and John M. Eversole. *The fire chief's handbook,* 6th ed. Dallas, TX: PennWell Books, 2003.

Bass, Bernard M. "Leadership: Good, better, best." *Organizational Dynamics* 13 (1985): 26–40.

Bass, Bernard M. "From transactional to transformational leadership: Learning to share the vision." *Organizational Dynamics* 18 (1990): 19–31.

Bass, Bernard M. "Concepts of leadership: The beginnings." In *The leader's companion*, edited by J. Thomas Wren, 49–52. New York: The Free Press, 1995.

Benest, Frank. *Marketing your budget*. Tampa, FL: The Innovation Groups, 1997.

Bennett, Lawrence T. *Fire service law*. Upper Saddle River, NJ: Prentice Hall, 2007.

Bennis, Warren, and Robert Thomas. "Crucibles of leadership." *Harvard Business Review* 80, no. 9 (2002): 39–45.

Bland, Richard E. *America burning: The report of the National Commission on Fire Prevention and Control*. Washington, DC: U.S. Department of Commerce, 1973.

Bland, Robert L., and Irene S. Rubin. *Budgeting: A guide for local government*. Washington, DC: International City/County Management Association, 1997.

Brunacini, Alan. *Fire command*, 2nd ed. Quincy, MA: National Fire Protection Association, 2002.

Buckman, John M., III, ed. *Chief fire officer's desk reference*. Sudbury, MA: Jones & Bartlett Learning, 2006.

Burlington Industries v. Ellerth, 524 U.S. 742 (1998).

Burlington Northern v. White, 548 U.S. 53 (2006).

Burn Institute. *Fire & burn prevention for seniors*. San Diego, CA: Burn Institute, 2010, http://www. burninstitute.org/fbp/programs/seniors.html.

Burns, James MacGregor. "Leadership." In *The leader's companion*, edited by J. Thomas Wren, 483. New York: The Free Press, 1995.

Camara v. San Francisco, 387 U.S. 523 (1967).

Carter, Harry R. "Being a good staff officer." *Fire Command* April (1985): 29.

Carter, Harry R. "Budget justification, your fight for a piece of the pie." *Fire Command* August (1986): 45.

Carter, Harry R. *Firefighting strategy and tactics*. Stillwater, OK: Fire Protection Publications, 1998.

Carter, Harry R. *Managing fire service finances*. Ashland, MA: International Society of Fire Service Instructors, 1989.

Carter, Harry R. *A master plan study of the Evesham Township Fire District*. Adelphia, NJ: Harry R. Carter, 1997.

Carter, Harry R., and Erwin Rausch. *Management in the fire service*, 4th ed. Sudbury, MA: Jones & Bartlett Learning, 2008.

Casimir, Gian. "Combinative aspects of leadership style: The ordering and temporal spacing of leadership behaviors." *Leadership Quarterly* 12 (2001): 245–278.

Chapman, Elwood N. *Life is an attitude*. Menlo Park, CA: Crisp Publications, 1992.

City of Ontario, CA v. Quon, 130 S.Ct. 2619, 560 U.S. 746 (2010).

Civil Rights Act. (1866).

Civil Rights Act of 1964. 42 USCA § 2000.

Clark, William F. *Firefighting principles and practices*. Saddle Brook, NJ: Fire Engineering Texts, 1991.

Cleveland Board of Education v. Loudermill, 470 U.S. 532 (1985).

Coleman, Ronny J. *Management of fire service operations*. Duxbury, MA: Wadsworth, 1978.

Commission on Fire Accreditation International. *Creating and evaluating standards of response coverage for fire departments*, 4th ed. Chantilly, VA: CFAI, 2008.

Compton, Dennis. *When in doubt, lead!* Stillwater, OK: Fire Protection Publications, 1999.

Compton, Dennis, and John Granito. *Managing fire and rescue services*. Washington, DC: International City Manager's Association, 2003.

Coolidge v. New Hampshire, 403 U.S. 443, 465–466 (1971).

Cote, Arthur. *Fire protection handbook*, 20th ed. Quincy, MA: National Fire Protection Association, 2007.

Cotter, John P. "What leaders really do." In *The leader's companion*, edited by J. Thomas Wren, 114–123. New York: The Free Press, 1995.

Covey, Stephen R. *The 7 habits of highly effective people*. New York: Franklin Covey Company, 1989.

Cronin, Thomas E. "Leadership and democracy." In *The leader's companion*, edited by J. Thomas Wren, 303–309. New York: The Free Press, 1995.

Daubert v. Merrell Dow Pharmaceuticals, 509 U.S. 579 (1993).

De Pree, Max. *Leadership is an art*. New York: Dell Publishing, 1989.

Digman, John M. "Personality Structure: Emergence of the Five-Factor Model." *Annu. Rev. Psychol.* 41 (1990): 417–440. www.annualreviews.org.

Ditzel, Paul C. *Fire engines, firefighters: The men, equipment, and machines, from colonial days to the present.* New York: Crown, 1976.

Dvir, Taly, Dov Eden, Bruce J. Avolio, and Boas Shamir. "Impact of transformational leadership on follower development and performance: A field experiment." *Academy of Management Journal* 45 (2002): 735–744.

Equal Employment Opportunity Commission, Department of Justice Right Division. *ADA Questions and Answers Publication*, 2002.

Eversole, John. *The fire chief's handbook*, 6th ed. Tulsa, OK: PennWell Corporation, 2003.

Faragher v. City of Boca Raton, 525 U.S. 775 (1998).

Favreau, Donald. *Fire service management*. New York: Fire Engineering Publications, 1973.

Federal Emergency Management Agency. "What is mitigation?," August 17, 2010, http://www.fema.gov/what-mitigation.

Federal Emergency Management Agency. "Lessons learned information sharing." (n.d.-a). https://www.llis.dhs.gov/index.do.

Federal Emergency Management Agency. "Preparedness: Continuous cycle." (n.d.-b). https://emilms.fema.gov/IS700aNEW/NIMS0102070.htm.

Federal Emergency Management Agency. *Residential structure and building fires, U.S. Fire Administration, October, 2008.* Emmitsburg, MD: U.S. Fire Administration, 2008.

Federal Whistleblower Act. 5 USC § 1201 et seq. (1989).

Fiedler, Fred. "The effect of inter-group competition on group member adjustment." *Personnel Psychology* 20 (1967): 33–44.

Fire Department Promotion Act. Illinois Compiled Statutes, Local Government (50 ILCS 742), 2003.

Foley v. Town of Randolph, 598 F.3d 1 (1st Cir. 2010).

Garcetti v. Ceballos, 547 U.S. 410 (2006).

Garrity v. New Jersey, 385 U.S. 493 (1967).

Gardner, John W. "The cry for leadership." In *The leader's companion*, edited by J. Thomas Wren, 3–7. New York: The Free Press, 1995.

George, Claude S. *Supervision in action: The art of managing others*, 4th ed. Reston, VA: Reston Publishing, 1985.

GE v. Joiner, 522 U.S. 136 (1997).

Graham, G. "Graham's rules for the elimination of civil liability." *GRECL*. Muskegon, MI: Gordon Graham, 2008.

Greenleaf, Robert K. "The servant as leader." In *Servant leadership—a journey into the nature of legitimate power and greatness*, edited by Larry C. Spears, 21–61. Mahwah, NJ: Paulist Press, 1977.

Hamm, Robert. *Leadership in the fire service*. Stillwater, OK: International Fire Service Training Association, 1990.

Hayward, Steven F. *Churchill on leadership*. Rocklin, CA: Prima Publishing, 1998.

Hersey, Paul, and Kenneth W. Blanchard. "Situational leadership." In *The leader's companion*, edited by J. Thomas Wren, 207–211. New York: The Free Press, 1995.

Howell, Jon P. "Substitutes for leadership: Their meaning and measurement—An historical assessment." *Leadership Quarterly* 8 (1997): 113–116.

Illinois Whistleblower Act. 740 ILCS (2004).

International Association of Fire Chiefs. *Officer development handbook*, 2nd ed. Fairfax, VA: IAFC, 2010.

International Code Council. *International Fire Code*. Country Club Hills, IL: ICC, 2006.

Irving, John. *My movie business: A memoir*. New York: Ballantine Books, 2000.

ISO. "ISO information about property/casualty insurance risk." *ISO's PPC Program*, 2011, http://www.iso.com/Research-and-Analyses/Studies-and-Whitepapers/ISO-s-PPC-Program-Page-2.html.

Jenaway, William F., and Daniel B. C. Gardiner. *Fire protection in the 21st century*. Ashland, MA: Alliance for Fire and Emergency Management, 1994.

Johnson Foundation. The *Wingspread Conference on Fire Service Administration, Education, and Research*. Racine, WI: Johnson Foundation, 1966.

Johnson Foundation. *Wingspread II: The Fire Problem in the United States*. Racine, WI: Johnson Foundation, 1976.

Johnson Foundation. *Wingspread III Conference on Fire Service Administration, Education, and Research*. Racine, WI: Johnson Foundation, 1986.

Johnson Foundation. *Wingspread IV Conference on Fire Service Administration, Education, and Research*. Racine, WI: Johnson Foundation, 1996.

Karter, Michael J., Jr., and Joseph L. Molis. *Firefighter injuries in the United States*. Quincy, MA: National Fire Protection Association, November 2014, http://www.nfpa.org/research/reports-and-statistics/the-fire-service/fatalities-and-injuries/firefighter-injuries-in-the-united-states.

Kerrigan, Heather. "Chicago's police misconduct cases go to court: To cut costs and save face, all of Chicago's police misconduct cases are going to trial instead of settling out of court." *Governing*, February 1, 2011, http://www.governing.com/topics/public-justicesafety/Chicagos-Police-Misconduct-Cases-Go-to-Court.html.

Kruse, Kevin. "What is leadership?" *Forbes*, April 9, 2013, http://www.forbes.com/sites/kevinkruse/2013/04/09/what-is-leadership/.

Kumho Tire Co. v. Carmichael, 526 U.S. 137 (1999).

Landy, Frank J., and Jeffrey M. Conte. *Work in the 21st century*, 2nd ed. Hoboken, NJ: Blackwell Publishing, 2007, 185–186.

Lane v. Franks, 134 S.Ct. 2374-75 (2014).

Layman, Lloyd. *Attacking and extinguishing interior fires*. Boston: National Fire Protection Association, 1955.

Ledbetter v. Goodyear Tire, 550 U.S. 532 (2007).

Lewis v. City of Chicago, 557 U.S. 08-974 (2010).

Lopes, Benjamin, F., III. "Office management and workflow." In *The fire chief's handbook*, 6th ed., edited by Robert C. Barr and John M. Eversole, 13–36. Dallas, TX: PennWell Books, 2003.

Marbury v. Madison, 5 U.S. 137 (1803).

Marinucci, Richard A. "Volunteer, paid on-call, and combination departments." In *The fire chief's handbook*, 6th ed., edited by Robert C. Barr and John M. Eversole, 925–945. Dallas, TX: PennWell Books, 2003.

Merrill v. Monticello, 138 U.S. 673 (1891).

Michael, Stephen R. "Doing what comes naturally." *Management Review* 65, no. 11 (1976): 20–31.

Michigan v. Clifford, 464 U.S. 287 (1984).

Michigan v. Tyler, 436 U.S. 499 (1978).

Michigan State Legislature. *Fire Prevention Code Act 207 of 1941*, 1978.

Minnesota Statutes. § 604.06 (2010).

Muskegon Area Medication Disposal Program. 2015. *About Us*. http://www.mamdp.com/ABOUT_MAMDP.html.

Nadler, David A., and Michael L. Tushman. "Beyond the charismatic leader: Leadership and organizational change." In *The leader's companion*, edited by J. Thomas Wren, 108–113. New York: The Free Press, 1995.

National Association of Counties. *Research brief: Dillon's Rule or not?* Washington, DC: NACO. Vol. 2, No. 1. 2004. http://www.celdf.org/downloads/Home%20Rule%20State%20or%20Dillons%20Rule%20State.pdf.

National Fire Protection Association. "About NFPA." (n.d.), http://www.nfpa.org/categoryList.asp?categoryID=143&URL=About%20NFPA.

National Fire Protection Association. *NFPA 101 Life safety code*. Quincy, MA: National Fire Protection Association, 2009.

National Fire Protection Association, International Association of Fire Chiefs, and International Society of Fire Service Instructors. *Fire service instructor: principles and practice*. Sudbury, MA: Jones & Bartlett Learning, 2009.

National Fire Protection Association. *NFPA 1: Fire code*. Quincy, MA: NFPA, 2015.

National Fire Protection Association. *NFPA 1021: Standard for fire officer professional qualifications*. Quincy, MA: NFPA, 2014.

National Fire Protection Association. "The Fire Service." Quincy, MA: National Fire Protection Association. http://www.nfpa.org/research/reports-and-statistics/the-fire-service.

National Treasury Employees Union v. Von Rabb, 489 U.S. 656 (1989).

National Volunteer Fire Council. *Retention and recruitment in the volunteer fire service*. Emmitsburg, MD: United States Fire Administration, 1993.

NLRB v. J. Weingarten, Inc., 420 U.S. 251 (1975).

O'Connor v. Ortega, 480 U.S. 709 (1987).

Paulsgrove, Robin. "Fire department administration and operations." In *Fire protection handbook*, 19th ed. Quincy, MA: National Fire Protection Association, 2003.

Peterson v. City of Mesa, 83 P.3d 35 (Ariz. 1994).

Peters, Thomas J., and Robert H. Waterman, Jr. *In search of excellence*. New York: Warner Books, 1982.

Pickering v. Board of Education, 31 U.S. 563 (1968).

Parry, John W. *Disability discrimination law, evidence and testimony*. Chicago: American Bar Association, 2009.

Ready.gov. "Make fire safety part of the plan." 2011, http://www.ready.gov/business/plan/planfiresafety.html.

Ricci v. DeStefano, 129 S.Ct. 2658 (2009).

Rost, Joseph C. "Leadership: A discussion about ethics." *Business Ethics Quarterly* 5, no. 1 (1995): 129–142.

See v. Seattle, 387 U.S. 541 (1967).

Schaeffer, Leonard D. "The leadership journey." *Harvard Business Review* 80, no. 10 (2002): 42–47.

Skinner v. Railway Labor Executives' Association, 489 U.S. 602 (1989).

Society for Human Resource Management (SHRM). "Retention: How do I calculate retention? Is retention related to turnover?" Alexandria, VA: Society for Human Resource Management. August 31, 2012, http://www.shrm.org/templatestools/hrqa/pages/calculatingretentionandturnover.aspx.

Society for Human Resource Management, n.d., Module 2, 2-16.

Sternberg, Robert J. "WICS: A model of leadership in organizations." *Academy of Management Learning and Education* 2 (2003): 386–401.

Summers v. Altarum Institute Corp., 740 F.3d 325 (2014).

Toyota Motor Manufacturing, Kentucky, Inc. v. Williams, 534 U.S. 184 (2002).

Turner, Nick, Julian Barling, Olga Epitropaki, Vicky Butcher, and Caroline Milner. "Transformational leadership and moral reasoning." *Journal of Applied Psychology* 87 (2002): 304–311.

U.S. Census Bureau. "State & County QuickFacts." Washington, DC: U.S. Census Bureau, 2010, http://quickfacts.census.gov/qfd/states/00000.html.

U.S. Constitution, Amendment XIV.

U.S. Department of the Army. *Military leadership*. FM 22-100. Ft. Benning, GA: U.S. Army, 1973.

U.S. EEOC v. City of St. Paul, 671 F.2d 1162, 1164 (8th Cir. 1982).

U.S. EEOC. *Questions and answers on the final rule implementing the ADA Amendments Act of 2008*. Washington, DC: EEOC, 2011.

U.S. Fire Administration. *Residential structure and building fires*. Emmitsburg, MD: U.S. Fire Administration, 2008.

U.S. Fire Administration, *Strategic analysis of community risk reduction*. Emmitsburg, MD: U.S. Fire Administration, 1994.

Utecht, Ronald E., and William D. Heier. "The contingency model and successful military training." *Academy of Management Journal* 19 (1976): 606–619.

Varone, J. Curtis. *Fire officer's legal handbook*. Clifton Park, NY: Delmar Thomson Learning, 2007.

Varone, J. Curtis. *Legal considerations for fire and emergency services*. Clifton Park, NY: Delmar Thomson Learning, 2006.

Volunteer and Combination Section of the International Association of Fire Chiefs. *A call for action: preserving and improving the future of the volunteer fire service*. Fairfax, VA: International Association of Fire Chiefs, 2004.

Volunteer and Combination Section of the International Association of Fire Chiefs. *Lighting the path of evolution: Leading the transition in volunteer and combination fire departments*. Fairfax, VA: International Association of Fire Chiefs, 2005.

Volunteer and Combination Section of the International Association of Fire Chiefs. *Managing the business of the fire department*. Fairfax, VA: International Association of Fire Chiefs, 2006.

Volunteer and Combination Section of the International Association of Fire Chiefs. *We're here for life: Leading and managing EMS in volunteer and combination fire departments*. Fairfax, VA: International Association of Fire Chiefs, 2008.

Von Clausewitz, Carl. *On war*. New York: Penguin Publishing, 1987.

Von Schell, Adolph. "Battle leadership." *Ft. Benning* (GA) *Herald*. (Orig. pub. 1933.) Reprinted by the Marine Corps Association, Quantico, VA, 1982.

Waite, Mitchell. *Fire service leadership: Theories and practices*. Burlington, MA: Jones & Bartlett Learning, 2008.

Wallace, Mark. *Fire department strategic planning: Creating future excellence*. Saddle Brook, NJ: Fire Engineering Books, 2006.

Windisch, Fred C., and Fred C. Crosby. *A leadership guide for combination fire departments*. Sudbury, MA: Jones & Bartlett Learning, 2008.

Wofford, Jerry C., J. Lee Whittington, and Vicki L. Goodwin. "Follower motive patterns as situational moderators for transformational leadership effectiveness." *Journal of Managerial Issues* 13 (2001): 196–211.

Wren, J. Thomas. *The leader's companion*. New York: The Free Press, 1995.

Young v. UPS, 135 S.Ct. 1338 (2015).

Yusko, Kenneth P., and Harold W. Goldstein. "Selecting and developing crisis-based leaders using competency-based simulations." *Journal of Contingencies and Crisis Management* 5 (1997): 216–223.

An Extract from NFPA 1021, *Standard for Fire Officer Professional Qualifications*, 2014 Edition

Chapter 6 Fire Officer III

6.1 General. For qualification at Fire Officer Level III, the Fire Officer II shall meet the job performance requirements defined in Sections 6.2 through 6.8 of this standard.

6.1.1* General Prerequisite Knowledge. Current national and international trends and developments related to fire service organization, management, and administrative principles, as well as public and private organizations that support the fire and emergency services and the functions of each.

6.1.2 General Prerequisite Skills. The ability to research, to use evaluative methods, to analyze data, to communicate orally and in writing, and to motivate members.

6.2 Human Resource Management. This duty involves establishing procedures for hiring, assigning, promoting, and encouraging professional development of members, according to the following job performance requirements.

6.2.1 Establish personnel assignments to maximize efficiency, given knowledge, training, and experience of the members available in accordance with policies and procedures, so that human resources are used in an effective manner.

(A) **Requisite Knowledge.** Minimum staffing requirements, available human resources, and policies and procedures.

(B) **Requisite Skills.** The ability to relate interpersonally and to communicate orally and in writing.

6.2.2 Develop procedures for hiring members, given policies of the AHJ and legal requirements, so that the process is valid and reliable.

(A) **Requisite Knowledge.** Applicable federal, state/provincial, and local laws; regulations and standards; and policies and procedures.

(B) **Requisite Skills.** The ability to communicate orally and in writing.

6.2.3 Develop procedures and programs for promoting members, given applicable policies and legal requirements, so that the process is valid and reliable, job-related, and nondiscriminatory.

(A) **Requisite Knowledge.** Applicable federal, state/provincial, and local laws; regulations and standards; and policies and procedures.

(B) **Requisite Skills.** The ability to communicate orally and in writing, to encourage professional development, and to mentor members.

6.2.4 Describe methods to facilitate and encourage members to participate in professional development, given a professional development model, so that members achieve their personal and professional goals.

(A) **Requisite Knowledge.** Interpersonal and motivational techniques, professional development model, goal setting, and personal and organizational goals.

(B) **Requisite Skills.** The ability to evaluate potential, to communicate orally, and to counsel members.

6.2.5 Develop a proposal for improving an employee benefit, given a need in the organization, so that adequate information is included to justify the requested benefit improvement.

(A) **Required Knowledge.** Agency's benefit program.

(B) **Required Skills.** The ability to conduct research and to communicate orally and in writing.

6.2.6 Develop a plan for providing an employee accommodation, given an employee need, the requirements, and applicable law, so that adequate information is included to justify the requested change(s).

(A) **Required Knowledge.** Agency's policies and procedures, and legal requirements or reasonable accommodations.

(B) **Required Skills.** The ability to conduct research and to communicate orally and in writing.

6.2.7 Develop an ongoing education training program, given organizational training requirements, so that members of the organization are given appropriate training to meet the mission of the organization.

(A) **Required Knowledge.** Agency mission and goals, training program development, and needs assessment.

(B) **Required Skills.** Ability to perform a needs assessment and to communicate orally and in writing.

6.3 Community and Government Relations. This duty involves developing programs that improve and expand service and build partnerships with the public, according to the following job performance requirements.

6.3.1 Develop a community risk reduction program, given risk assessment data, so that program outcomes are met.

(A) **Requisite Knowledge.** Community demographics, resource availability, community needs, customer service principles, and program development.

(B) **Requisite Skills.** The ability to relate interpersonally, to communicate orally and in writing, and to analyze and interpret data.

6.4 Administration. This duty involves preparing a divisional or departmental budget, developing a budget management system, soliciting bids, planning for resource allocation, and working with records management systems, according to the following job performance requirements.

6.4.1 Develop a divisional or departmental budget, given schedules and guidelines concerning its preparation, so that capital, operating, and personnel costs are determined and justified.

(A) **Requisite Knowledge.** The supplies and equipment necessary for existing and new programs; repairs to existing facilities; new equipment, apparatus maintenance, and personnel costs; and approved budgeting system.

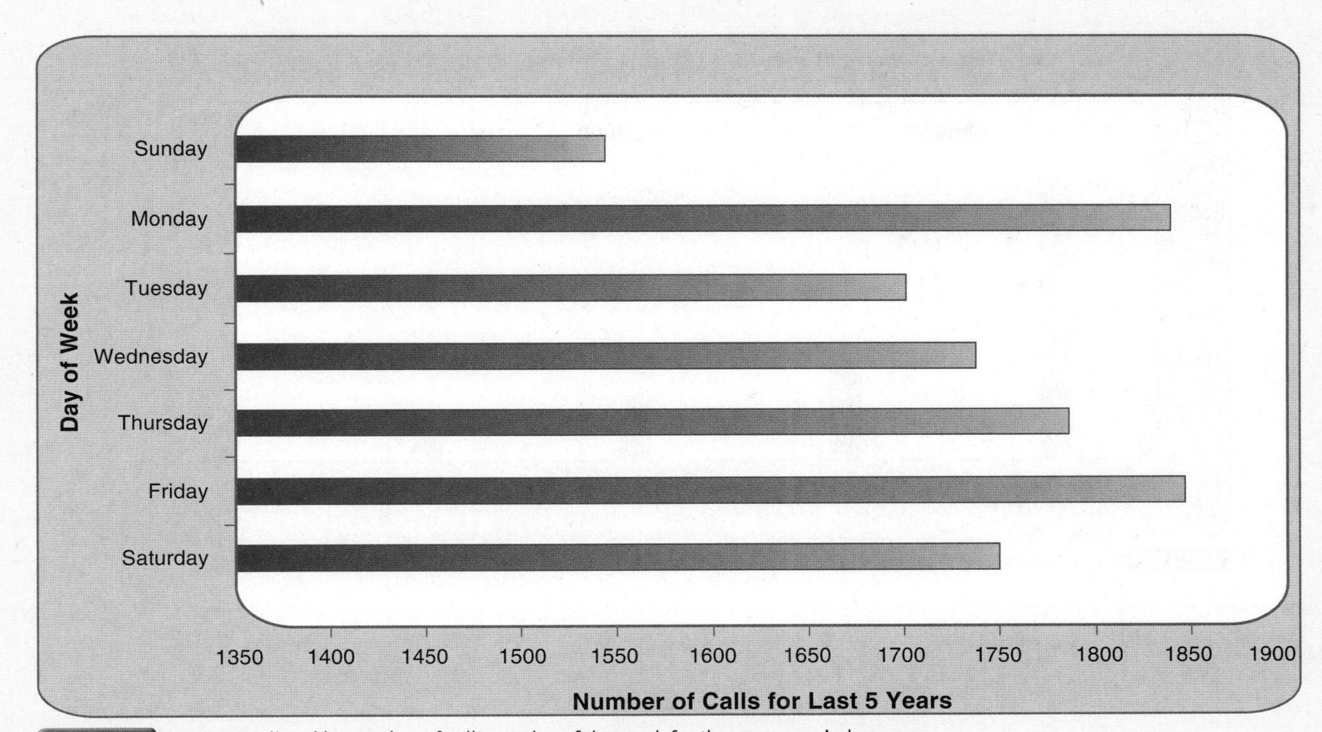

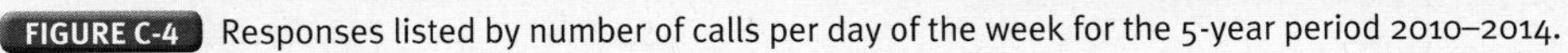
FIGURE C-4 Responses listed by number of calls per day of the week for the 5-year period 2010–2014.

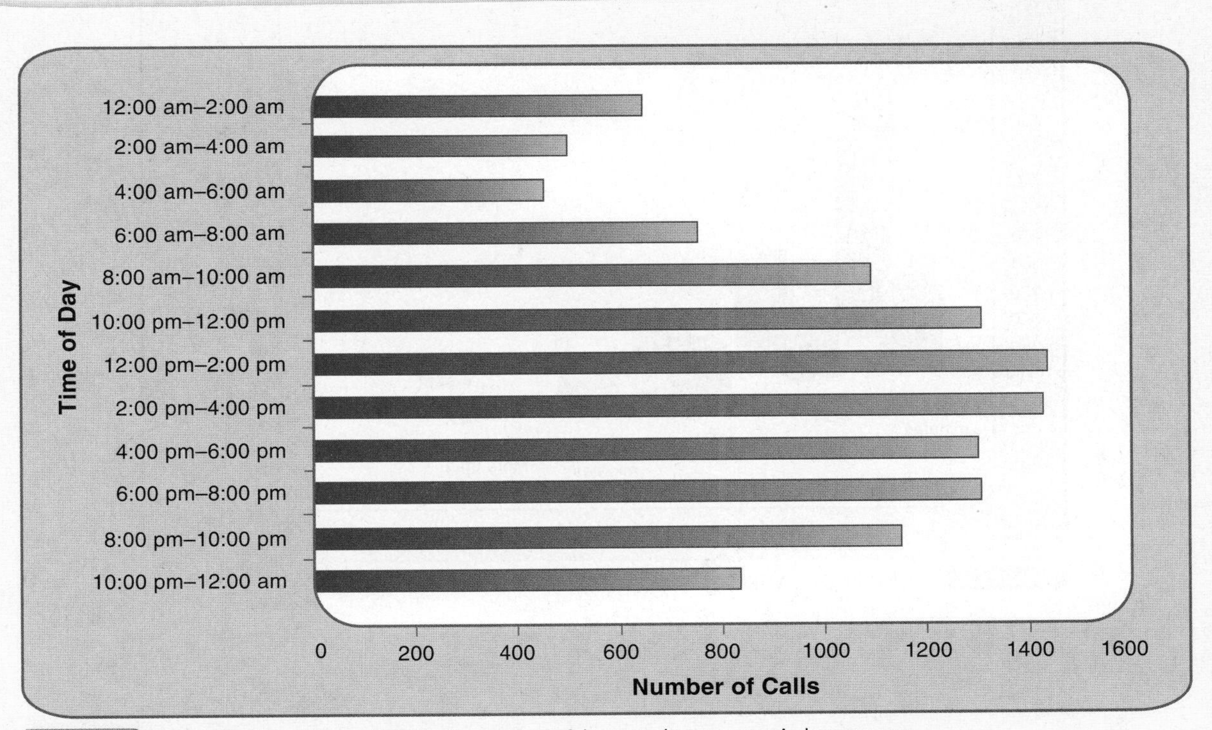

FIGURE C-5 Responses listed by number of calls per time of day over the 5-year period 2010–2014.

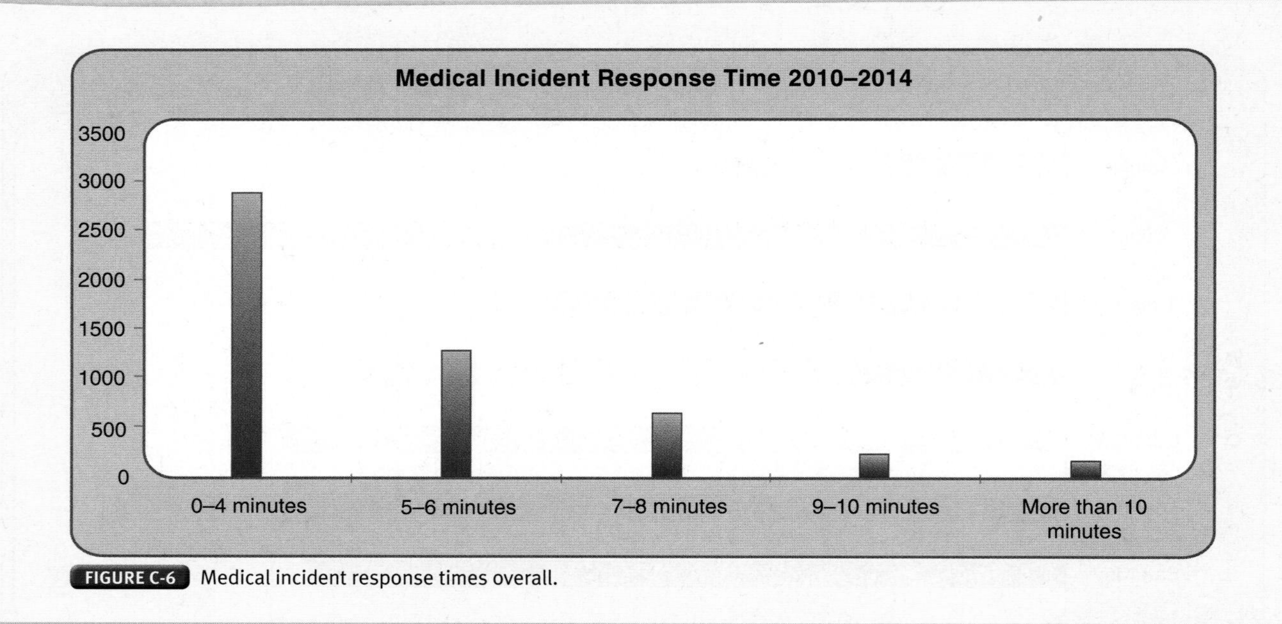

FIGURE C-6 Medical incident response times overall.

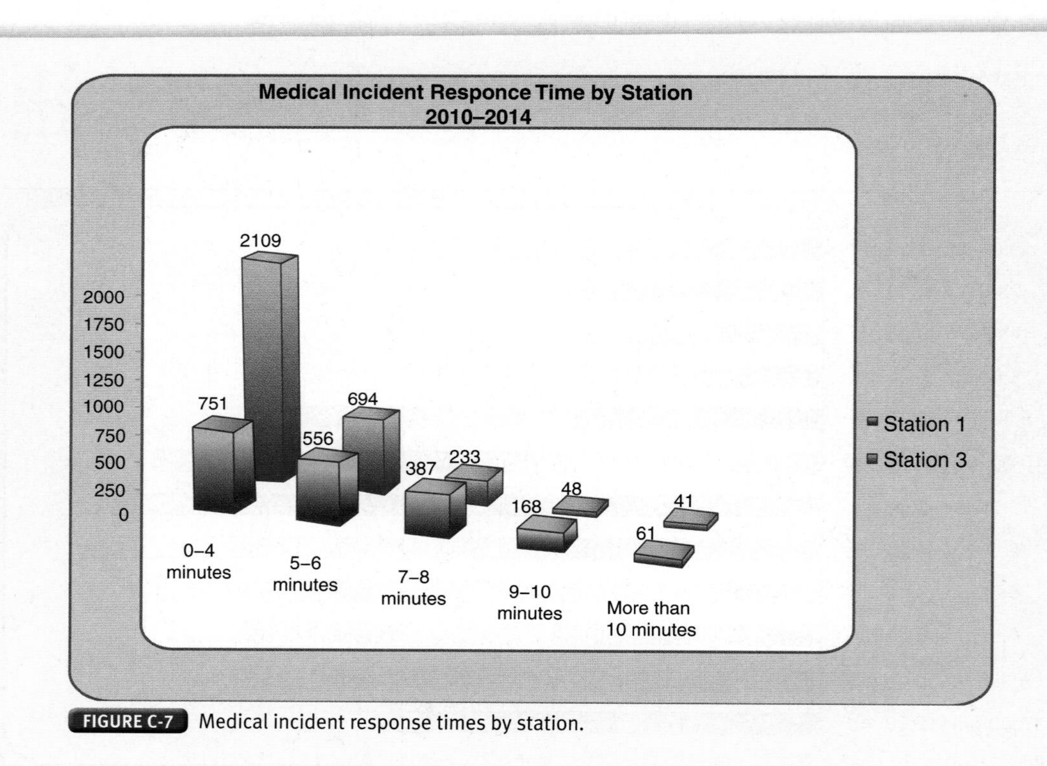

FIGURE C-7 Medical incident response times by station.

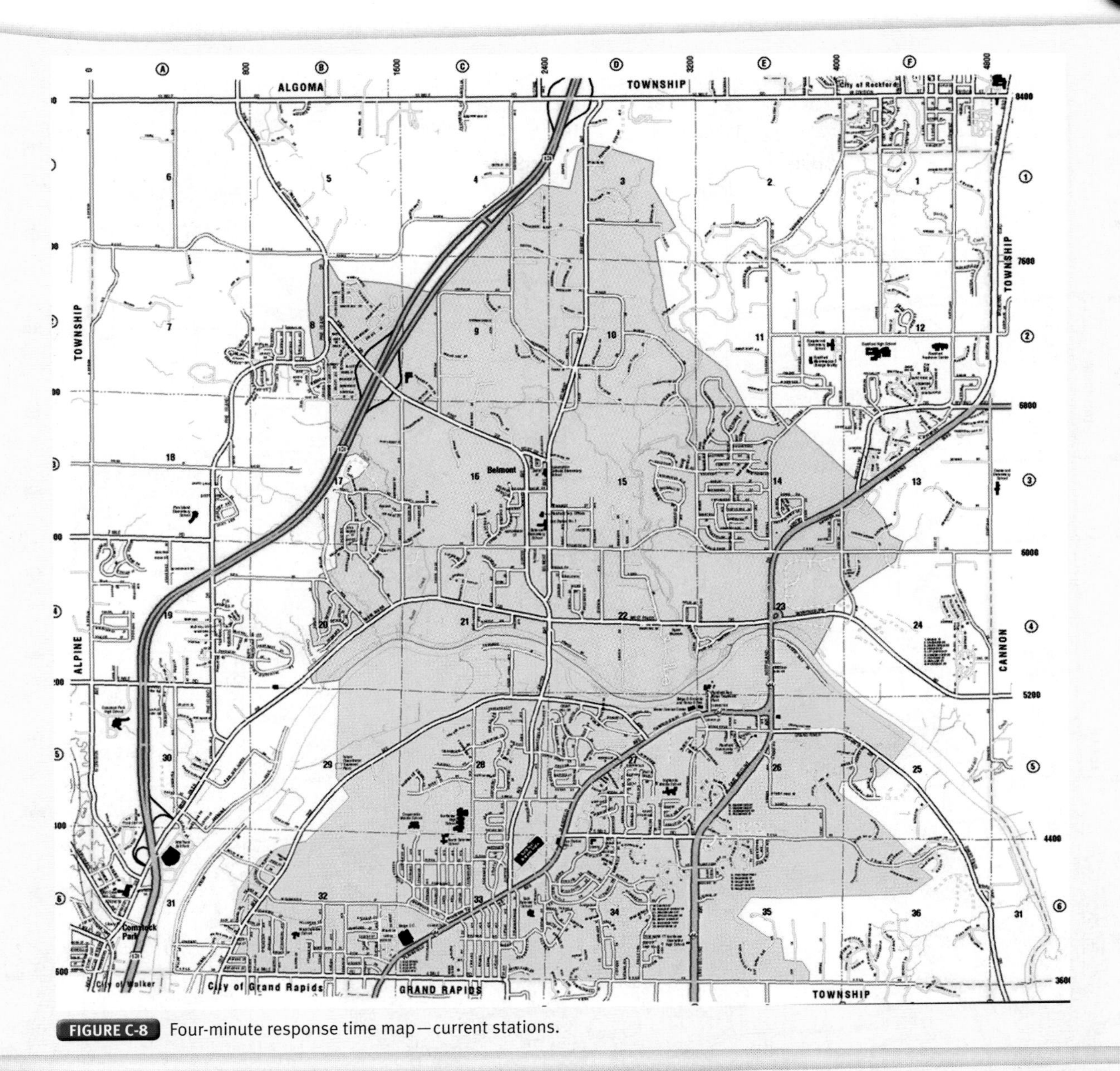

FIGURE C-8 Four-minute response time map—current stations.

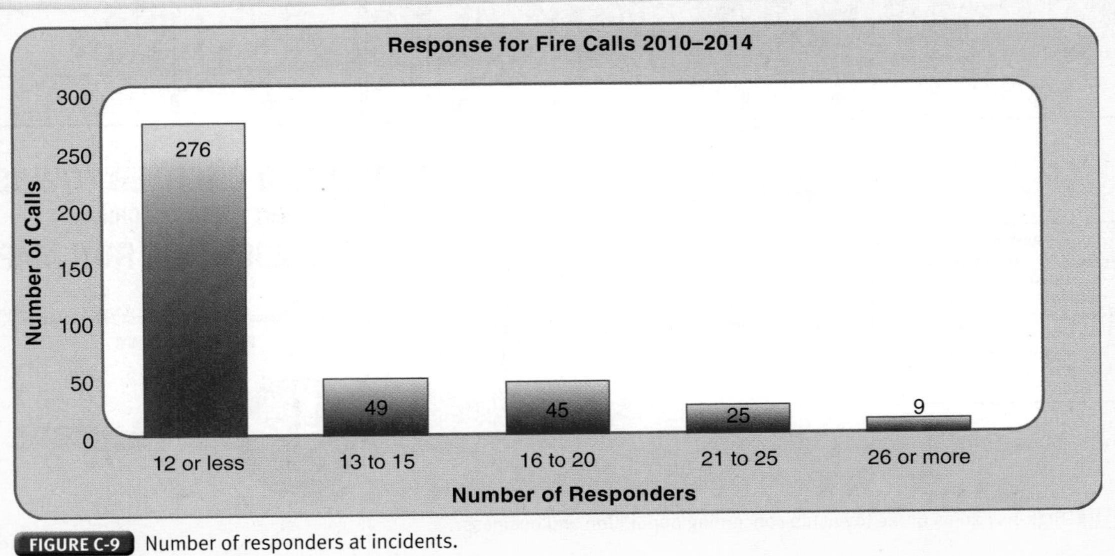

FIGURE C-9 Number of responders at incidents.

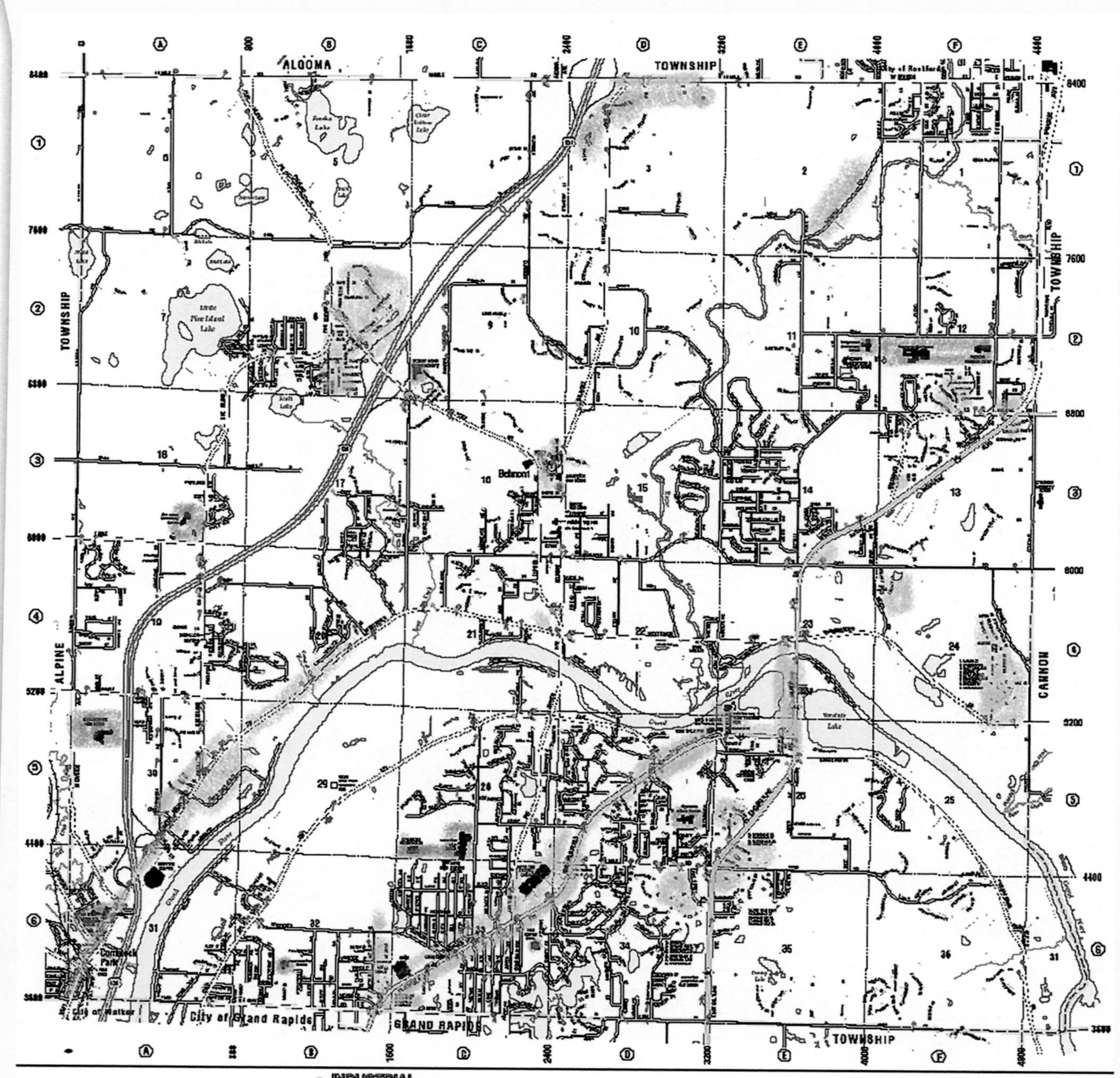

INDUSTRIAL
COMMERCIAL
HIGH DENSITY RESIDENTIAL
EDUCATIONAL

PLAINFIELD CHARTER TOWNSHIP

KENT COUNTY, MICHIGAN

2003-2006 FIRE RUN MAP

MAP LEGEND

Railroads
Rivers & Streams
Section Lines
Existing Buildings
Proposed Buildings
Governmental Units
Lakes & Ponds

Roads & Streets
Freeway
Freeway Ramp
Highway
Primary
Secondary
Proposed/Under Construction

FIRE RUN LEGEND

2003
2004
2005
2006

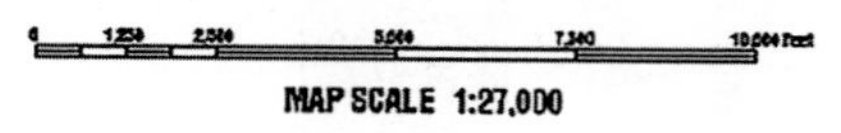

MAP SCALE 1:27,000

FIGURE C-10 High-risk areas of the township concerning population and business.

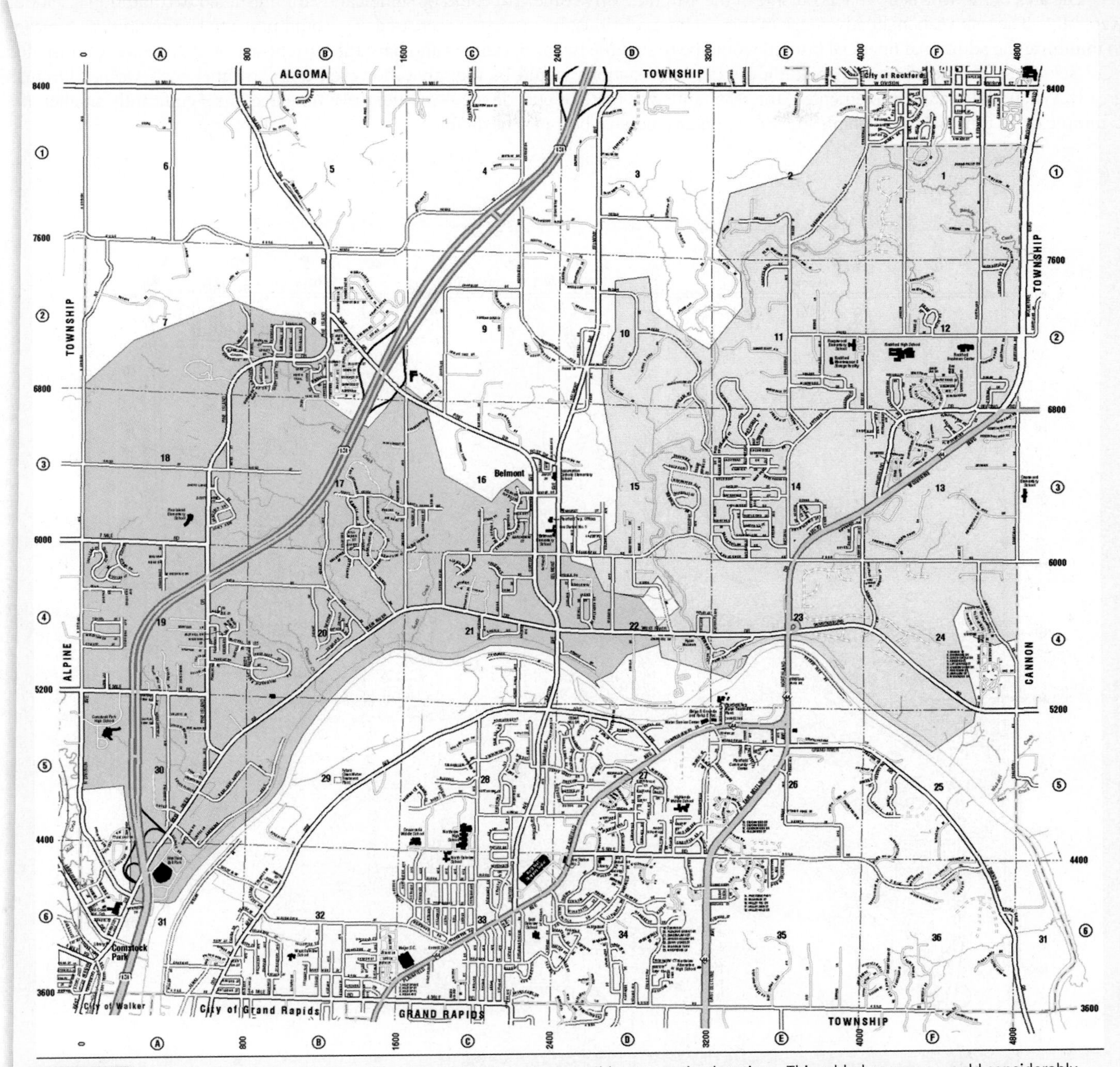

FIGURE C-11 This map defines the 4-minute drive time district from two possible new station locations. This added coverage would considerably improve the response capability. One station is located at the terminus of Brewer at Kuttshill; the second location used was the Buth tank site on Buth, east of Pine Island. The Buth site may not be the optimal location but is on Township-owned property and is in the proximity of where a station could/should/would be positioned.

The area of the Township that is outside of the 4-minute drive time goal could be significantly diminished if two additional stations were built. The risk vs. benefit advantage of this becomes obvious; however, the increased cost could prove to be prohibitive. A means to minimize the additional financial burden would be to staff the two new stations and have the current station #1 become an unstaffed POC station. The original cost of building the two new stations would be the same; however, the personnel costs would diminish greatly. When that plan is implemented, the resultant areas that are outside the 4-minute drive time become significantly smaller. If a 5-minute drive time were acceptable, then the noncovered area becomes negligible.

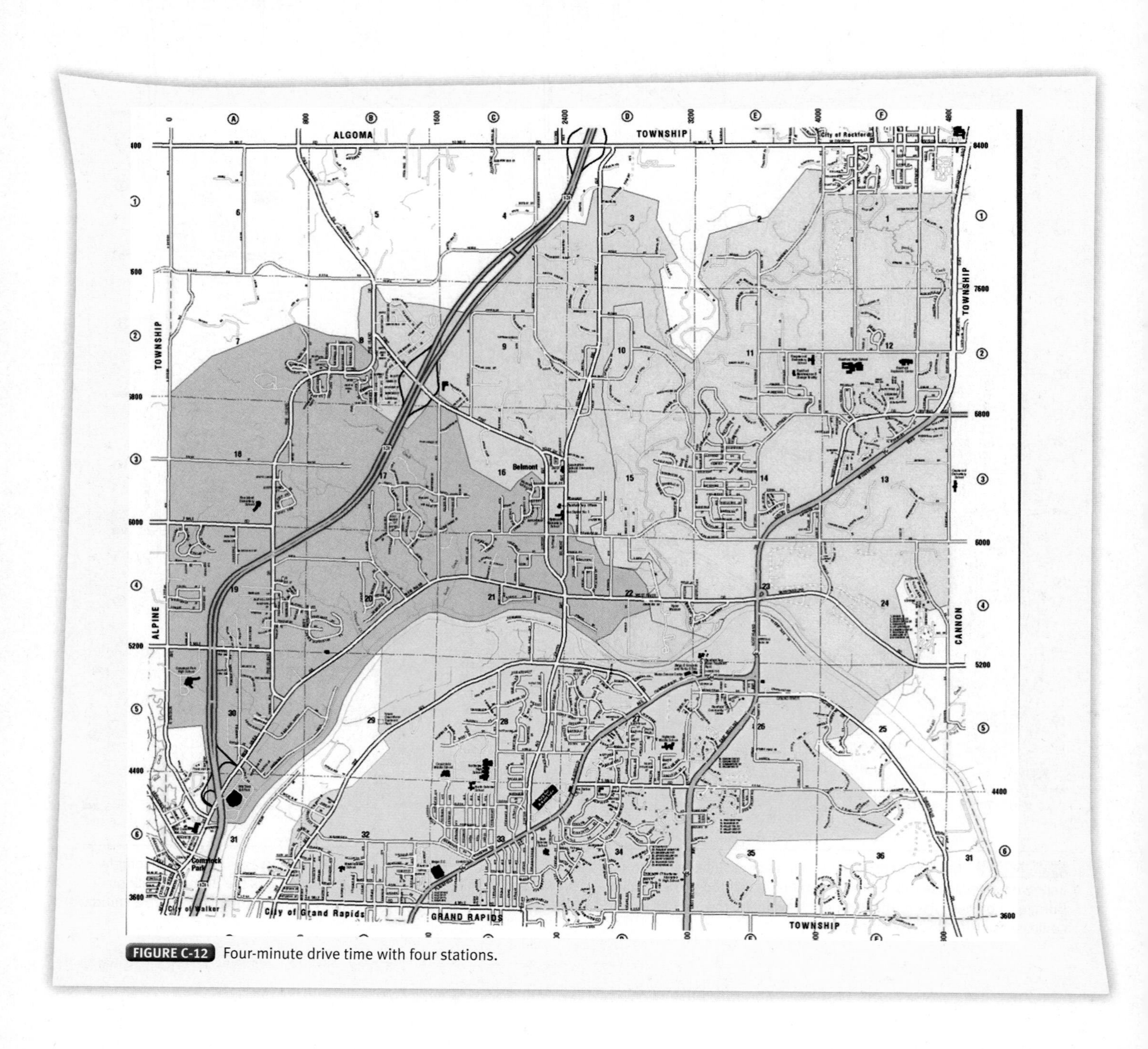

FIGURE C-12 Four-minute drive time with four stations.

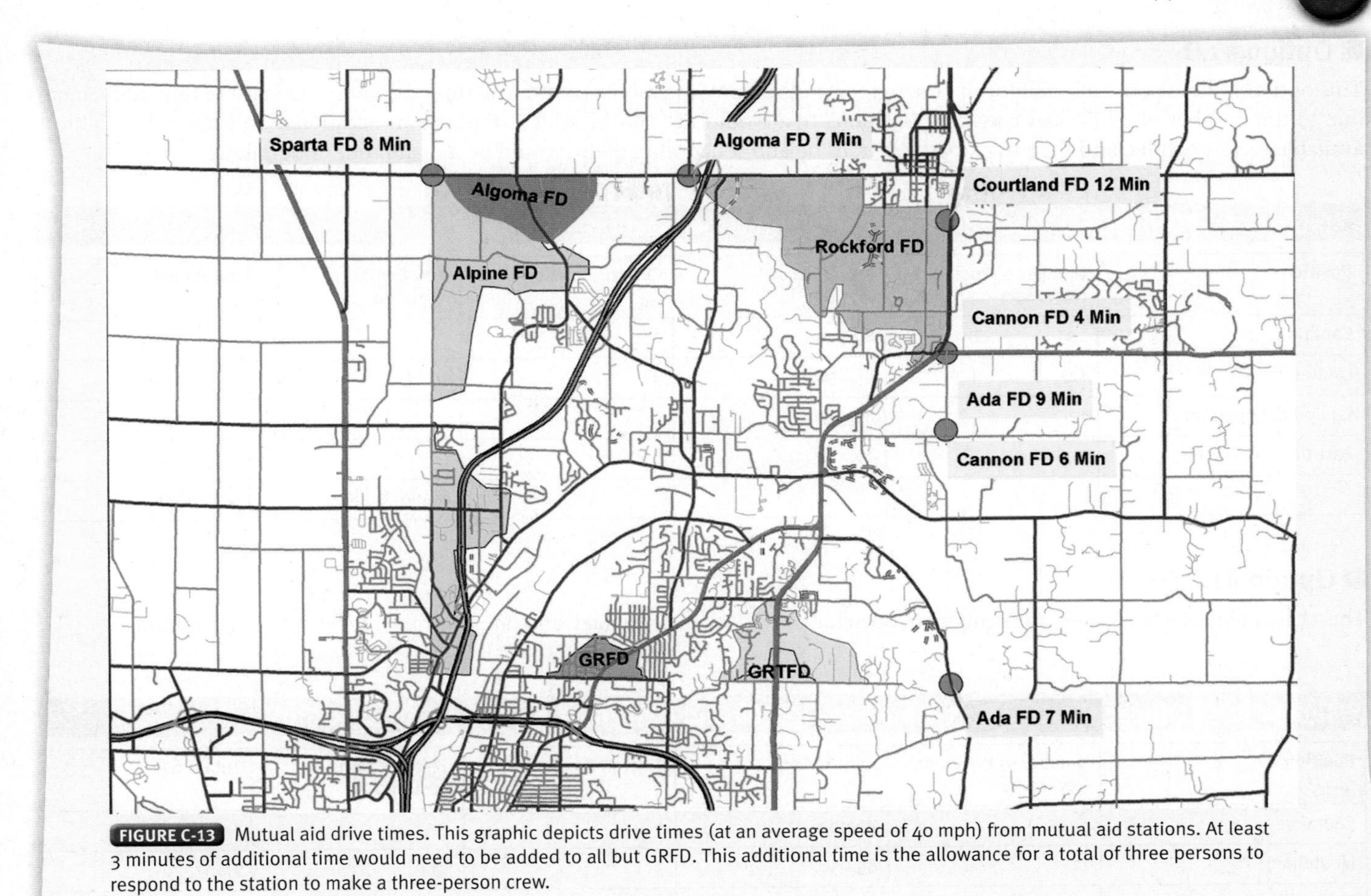

FIGURE C-13 Mutual aid drive times. This graphic depicts drive times (at an average speed of 40 mph) from mutual aid stations. At least 3 minutes of additional time would need to be added to all but GRFD. This additional time is the allowance for a total of three persons to respond to the station to make a three-person crew.

Staffing Options and Facility Information

As mentioned in the body of the report, there are two issues over which the fire department has direct control that affect their response to an emergency: station placement and staffing. The information in this appendix reflects the possibilities for providing an improved level of service through the addition of stations and staff and the cost of implementing each plan. The cost estimates for the stations are based on a cost-per-square-foot basis and do not reflect any actual quotes. The numbers reflected for the personnel expense are estimated on the projected 2007 wage rates *including all benefits* (they are not the actual wages) for all positions but captain. These wages have been calculated at the top of the scale and would be considerably lower the first year, escalating until the top is reached at 5 years (according to the wage used here). There is also a 54-hour overtime allowance built in. The wage for captain was projected at the lieutenant's rate +5%.

The current budget for personnel wages and benefits is $907,487.

Option #1 A

This is the most complete coverage option. This option requires four stations staffed 24/7 with four personnel and one shift supervisor. It would fully comply with NFPA 1710 and provide 12 fire fighters on scene within 12 minutes reflex time 90% of the time within the Township. It would leave one station available to respond to additional calls for service. The cost of this option is determined as follows:

Table C-5 Officer, EO, and Two Fire Fighters per Station, Four Stations Staffed

Position	Number of Positions Necessary	Annual Cost per Position	Current Positions	New Positions Required	Total Cost
Captain	3	89,920	0	3	269,760
Lieutenant	12	85,638	3	9	1,027,656
Equipment operator	12	80,197	7	5	962,364
Fire fighter	24	75,096	0	24	1,802,304
				Grand Total	4,062,084*

*In order to accurately apply this figure, the total amount budgeted for part-time fire fighters (226,326) has been subtracted from the original total.

Option #1 B

This option utilizes part-time staff for the fire fighter positions. While more economical, this option may be hard to fully implement due to the number of additional part-time positions that would need to be added. Finding an additional 26 trained fire fighters available to work shifts, and then tracking their training and scheduling them, would be a mammoth undertaking.

Table C-6 Officer, EO, Two Part-Time Fire Fighters per Station

Position	Number of Positions Necessary	Annual Cost	Current Positions	New Positions Required	Total Cost
Captain	3	89,920	0	3	269,760
Lieutenant	12	85,638	3	9	1,027,656
Equipment operator	12	80,197	7	5	962,364
Part-time fire fighter	24	37,721	6	18	905,304
				Grand Total	3,165,084

Option #1 C

This option uses one lieutenant, one equipment operator, one full-time fire fighter, and one part-time fire fighter at each station.

Table C-7 Officer, EO, One Part-Time Fire Fighter, and One Full-Time Fire Fighter per Station

Position	Number of Positions Necessary	Annual Cost	Current Positions	New Positions Required	Total Cost
Captain	3	89,920	0	3	269,760
Lieutenant	12	85,638	3	9	1,027,656
Equipment operator	12	80,197	7	5	962,364
Fire fighter	12	75,096	0	12	901,152
Part-time fire fighter	12	37,721	6	6	452,652
				Grand Total	3,613,584

Option #1 D

This option drops the number of fire fighters to one rather than two (a staff total of three as opposed to four), which requires that we have one position per shift to cover for the EO.

Table C-8 Officer, EO, and One Part-Time Fire Fighter per Station

Position	Number of Positions Necessary	Annual Cost	Current Positions	New Positions Required	Total Cost
Captain	3	89,920	0	3	269,760
Lieutenant	12	85,638	3	9	1,027,656
Equipment operator	12	80,197	7	5	962,364
Part-time fire fighter	12	37,721	6	6	452,652
				Grand Total	2,712,432

Option #1 E

This option provides no change from current staffing. The total may not seem to reconcile with current budget numbers; however, it reflects anticipated raises and is a worst-case scenario from a cost-per-employee standpoint (i.e., everyone is at the top of their salary range and we are including built in, and training overtime).

Table C-9 Worst Case Scenario for Cost per Employee with No Staffing Change

Position	Number of Positions Necessary	Annual Cost	Current Positions	New Positions Required	Total Cost
Captain	0	89,920	0	0	0
Lieutenant	3	85,638	3	0	256,914
Equipment operator	12	80,197	7	0	962,364
Part-time fire fighter	12	37,721	6	6	452,652
				Grand Total	1,671,930

■ Option #2 A

In this option, three stations would be used rather than four, two of which would be new, and the existing station #1 would consist of paid-on-call personnel. This option uses full-time fire fighters to fill all shifts.

Table C-10 Three Stations, Paid-on-Call and Full-Time Fire Fighters

Position	Number of Positions Necessary	Annual Cost	Current Positions	New Positions Required	Total Additional Cost
Captain	3	89,920	0	3	269,760
Lieutenant	9	85,638	3	6	770,742
Equipment operator	9	80,197	7	2	721,773
Fire fighter	18	75,096	0	18	1,351,728
				Grand Total	3,114,003*

■ Option #2 B

In this option, three stations would be used rather than four, two of which would be new, and the existing station #1 would consist of paid-on-call personnel. This option uses part-time fire fighters to fill shifts.

Table C-11 Three Stations, Paid-on-Call and Part-Time Fire Fighters

Position	Number of Positions Necessary	Annual Cost	Current Positions	New Positions Required	Total Cost
Captain	3	89,920	0	3	269,760
Lieutenant	9	85,638	3	6	770,742
Equipment operator	9	80,197	7	2	721,773
Part-time fire fighter	18	37,721	6	12	678,978
				Grand Total	2,441,253

■ Option #3 A

In this option, three stations would be used with the use of the existing station #1. Full-time fire fighters would fill all shifts.

Table C-12 Three Stations, Full-Time Fire Fighters Only

Position	Number of Positions Necessary	Annual Cost	Current Positions	New Positions Required	Total Cost
Captain	3	89,920	0	3	269,760
Lieutenant	9	85,638	3	6	770,742
Equipment operator	9	80,197	7	2	721,773
Fire fighter	18	37,721	0	18	1,351,728
				Grand Total	3,114,003*

*In order to accurately apply this figure, the total amount budgeted for part-time firefighters (226,326) has been subtracted from the original total.

D

G

H

I

J

K

L

M

P